RENAISSANCE GENIUS

Galileo Galilei & His Legacy to Modern Science

RENAISSANCE
GENIUS

Galileo Galilei & His Legacy to Modern Science

DAVID WHITEHOUSE

STERLING

New York / London
www.sterlingpublishing.com

STERLING and the distinctive Sterling logo are
registered trademarks of Sterling Publishing Co., Inc.

10 9 8 7 6 5 4 3 2 1

QUI PLANTAVIT CURABIT

A SPRINGWOOD BOOK

Published by Sterling Publishing Co., Inc.
387 Park Avenue South, New York, NY 10016

© 2009 by Springwood SA, Lugano, Switzerland

Text © 2009 by David Whitehouse, PhD

Art Acknowledgments and copyrights, see pp. 224-233

Distributed in Canada by Sterling Publishing
c/o Canadian Manda Group, 165 Dufferin Street
Toronto, Ontario, Canada M6K 3H6
Distributed in the United Kingdom by GMC Distribution Services
Castle Place, 166 High Street, Lewes, East Sussex, England BN7 1XU
Distributed in Australia by Capricorn Link (Australia) Pty. Ltd.
P.O. Box 704, Windsor, NSW 2756, Australia

Sterling ISBN 978-1-4027-6977-1

Book design and layout: Stefan Morris *stefan.morris@live.com*

For information about custom editions, special sales, premium and
corporate purchases, please contact Sterling Special Sales Department
at 800-805-5489 or specialsales@sterlingpublishing.com

TABLE OF CONTENTS

CHAPTER ONE

MEDIEVAL MADRIGALS

Galileo's father, Vincenzo, was a composer, a music theorist, and
a seminal figure in the musical life of the late Renaissance. His
discoveries in acoustics and in the physics of vibrating strings deeply
influenced Galileo. Vincenzo directed his son away from the pure
mathematical formulas followed at the time, toward experimentation
and observation, with results explained by
empiricist mathematics.

At the end of his life in 1633, under house arrest in his villa at Arcetri, bedridden and blind, Galileo wrote that his universe, the one whose boundaries he had extended a thousand fold, had shriveled up to become as narrow as a compass and could be filled by his own sensations. In his contemplation he turned to his lifelong love—the lute— and thought of his father, Vincenzo. Were it not for him Galileo would have become a priest or a painter, and perhaps one who would not have been remembered. He recalled the events that had confined him at Arcetri: his trial before the Inquisition when he had defended what he had seen with his "optik tube" against the accusation of heresy. A conflict with the Church was not what he had wanted. Despite his belligerent personality, his clashes with the authorities were brought about by others attacking him and his findings, not the other way round. Through it all, he had remained a devout Catholic. His two daughters had become nuns. He could have left for the Protestant north at any time, where he would have been safe and lauded. However, he had never intended to leave Italy. It was hard after so many years of being lionized by Rome, after his books had passed the approval of the papal censors, to endure this punishment: convicted by the Inquisition on a technicality with a sentence decided before the trial had even begun.

He thought of happier times, of the tower of Saint Mark that he and the rulers of Venice had climbed all those years ago with

Below *Portrait of Galileo Galilei.*

Above *Galileo's villa at Arcetri where he lived for many years.*

Above *Galileo Galilei and Vincenzo Viviani
discussing astronomy.*

his newly made telescope or *perspiculum*, as he had once called it. They pointed it out to the sea and saw people on nearby islands, and observed a ship for a full three hours before it was visible to the unaided eye. Everyone agreed it was a marvel. He remembered the first time he had turned his new instrument toward the sky, capturing the quivering moon in its narrow field of vision: the dappled moon, streaked with gray and silver, pockmarked with the tiny shadows of mountains and craters. He observed the sun, too, with its tiny spots that moved from day to day, and Jupiter and its circling moons.

There was so much to remember; the swinging pendulum, the falling weights from the leaning tower of Pisa, the thermometer, and the balance he had constructed. His thoughts always returned to the courtroom in the Holy Office. They had called him foolish and absurd and a "suspect of heresy." Then there were the implications of a harsher treatment if he did not recant his theories. His books were banned and he was silenced and imprisoned in his home. A portrait painted of him five years before his death shows the strain on his face and perhaps the pain in his eyes, so different from the earlier portraits that displayed a jaunty inquisitiveness and unshakable inner confidence. The bravado had gone but not his confidence in science. No matter what the Church decreed, he knew that science had been set free and would no longer be guided by philosophy. The imprisonment of his body and mind were but a temporary victory. Science would flourish from Galileo's time on but it was more than 350 years before the Church reflected on what it had done to him.

He had been blind for five years and his heart was failing but in his mind's eye, he could see the moon and Jupiter and the myriad stars. Days before his end Pope Urban VIII sent him his special blessing.

And so the dying scientist fingered his lute, attempting, perhaps for the last time, the madrigals that his father had taught him when he was ten years old, and knowing now that the very spacing of those frets and the position of his hands had been laid out by his father. He remembered the musical experiment they conducted together that changed not only the course of musical history but also of science. For without his father the world would never have heard much of Galileo Galilei, held by many to be one of the most important scientists who ever lived, indeed, the "father of modern science."

Above *Pope Urban VIII (1568–1644), born Maffeo Barberini, was the last pope to expand the papal territory by force of arms. Though he was the cause for Galileo's accusation of heresy by the Holy Inquisition, the pope sent him his special blessing on the last days of his life.*

Vincenzo Galilei was born in 1520 in the small hilltop village of Santa Maria a Monte in Tuscany, surrounded by olive groves and the gentle, wooded hills known as the Cerbaie. His family could be traced back many generations: his ancestors had been prominent members of the Florentine community, and nineteen previous Galileis had been members of the ruling body—the Signoria. They once had great wealth but that had long gone by Vincenzo's time. His father made his living in the wool trade.

When Vincenzo was a child, he learned the keyboard and the lute in the medieval style that he later rejected. One can still listen to his music today on an MP3 player. It sounds delicate and subdued, like the hills of the Cerbaie perhaps, but in its day, it was revolutionary and in its arpeggios can be found the source of his son Galileo's personality, originality and ultimately, of his conflict.

While still young, Vincenzo attracted the attention of wealthy and influential patrons of music, such as Bernardetto de' Medici, Prince of Ottaiano (died after 1576), and Giovanni Bardi, Count of Vernio (1534–1612). With Bardi's support, he was able to devote himself to the theory of music. He studied with the famous musical theorist Gioseffo Zarlino (1517–1590) in Venice, and traveled abroad to learn about the music of the Turks and Moors. As he traveled through Europe, the continent was changing as the Renaissance stirred man's gaze away from the obscurant mists of the medieval mindset that had fettered it for centuries.

Below *Bust of Girolamo Savonarola in profile. The Dominican priest would be remembered by history as the strongest opponent to the Renaissance.*

Italy, too, was changing. For many, the rise to power of the austere monk Girolamo Savonarola in Florence from 1494 to 1498 marked the end of the city's flourishing as the seat of the freethinking Renaissance. Savonarola preached against moral corruption and lack of adherence to scripture, even by the Borgia Pope; for others this time was marked by the triumphant return of the Medici to rule. However, when the Medici returned to power it was clear that the centre of the Renaissance was shifting from Florence to Rome and a reaction against intellectual freedom

Above *A mystical wheel. At the time Galileo was born, the medieval mindset that saw all the elements of the world functioning within a set divine order was still predominant.*

emerged with this change. In 1542, the Sacred Congregation of the Inquisition was formed, and a few years later its *Index Librorum Prohibitorum* banned a wide range of Renaissance works of literature. The scholarly freedom of the Renaissance was evaporating, as Rome became the arbiter of acceptable knowledge. Some saw the medieval mists returning.

As Vincenzo traveled to learn about music, a seventy-year-old priest lay dying in Frombork, Prussia, in today's modern Poland. Legend has it that the first printed copy of his book, *De Revolutionibus*, was thrust into the hands of Nicolaus Copernicus (1473–1543), a neighbor in Frombork, on the day of the priest's death. The book showed an alternative model of the universe, different from that endorsed by Scripture but based on observation. The earth was not the center of the universe as the Bible said — the sun was. At first, the book received little attention; there were no fierce sermons about it contradicting Holy Scripture. It was to take sixty years before Copernicus' work came under serious attack and was placed on the index of forbidden books by Rome. The reason why it became so controversial was due to Vincenzo's scientist son.

We owe our present-day knowledge of Greek music and its influence on the medieval musical landscape to Vincenzo's correspondence over ten years with musician Girolamo Mei (1519–1594), the author of a treatise on ancient music. Mei's writings held that music was a devotion to God and not a vehicle for mankind's creativity. At the time, only the theory and the philosophy of music were of paramount importance, and musicians played according to strict rules. It was unthinkable that a musician, who was after all only an artisan, would break the divine laws of mathematical harmony. However, Vincenzo did the unthinkable.

Throughout learned medieval Europe music had an equal place alongside arithmetic, geometry, and astronomy. The demand for music was great, as every court, whether regal or belonging to a provincial nobleman, had its musicians. Vincenzo became a respected teacher and musician and moved to Pisa, where his sons, among them Galileo, were born.

Above *Discussion between Theologian and Astronomer, from Alliaco's Concordatia Astronomiae cum Theologia, Erhat Ratdolt, Augsburg, 1490.*

Pisa was home to nine thousand poor souls who lived in a dank and uninviting backwater surrounded by malarial swamps. The two rivers at whose conjunction the city was built–the Arno and the Serchio—were fetid with sewage and silted up. Few outsiders came to Pisa because of its hostile reputation, but the Duke of Florence, Cosimo de' Medici, had plans for the city and by 1555 had begun draining the surrounding marshes and improving the sewers. He built new public buildings and spaces, unclogged the canals, and improved the living conditions of the Pisans. They responded positively: Trade increased as more boats used the river and the city's university emerged from decades of academic slumber.

Vincenzo lived a short distance outside the ancient wall of the city, in a quarter called San Francesco of artisans and shopkeepers. He had a small house near the market and made a modest living teaching music at the home of a noble family called Bocca. His income, however, was not sufficient to make ends meet and he traded in wool, which, unlike music, provided reliable revenue. His wife, Giulia Ammananti, did not feel Vincenzo was keeping her in the manner she deserved. Her family had been important, descendants of a Roman cardinal, and she never let her husband forget it.

Their eldest child, Galileo, was born on February 15, 1564, the first of seven children. He grew up in a household full of music and, most importantly, full of questions and a disrespect for authority that he inherited from his father, who was convinced that the medieval course of music was wrong in that it set about to favor mathematics and the philosophy of music above the harmonies sensed by the ear. Vincenzo questioned the accepted wisdom of Pythagoras and his ideas of propositions, octaves, and ratios in musical theory. Music, he believed, needed to the result of experiment, perception, and observation. He wanted to take the Pythagorean magic numbers out of music and turn the established order upside-down.

In 1572 when Galileo was eight years old, Vincenzo left his family behind and returned to his native Florence to work for a Florentine Count. The family joined him two years later and young Galileo began to learn the lute and watch his father experiment with new ways of playing music. His son's potential was evident to Vincenzo: Galileo had mastered his first instrument and in many ways equaled his father's

Left Though Galileo and his father lived in the Renaissance, many medieval beliefs were still held: for many as well as for the Curch the world was ruled by divine power.

Above A coin portraying Pythagoras, whose experiments of determining musical pitch and interval, Vincenzo and Galileo corrected.

playing. It was time, he decided, for his son to go away to study.

Vincenzo chose the Vallombrosa Monastery, famous for its discipline and academic study. There Galileo learned Latin, Greek, mathematics, and some science. When he showed signs of wanting to take holy orders, Vincenzo was horrified. He did not envisage his son becoming a priest. One day, with no notification, he arrived at the monastery and on the pretext of looking after his son's health took Galileo away. His mother had just given birth to her sixth child, and with so little space in the small house, it was decided that Galileo would return to Pisa to live with his cousin and learn the wool trade. For several years young Galileo half-heartedly traded in wool, but his ambition was set on entering the University of Pisa, and in the summer of 1581 he enrolled in the university to study medicine, a path once again chosen by his father.

Galileo was aimless in his studies, and the rigid rules of medieval medicine and its superstitions quickly led to disenchantment. He earned a reputation for contradicting his professors. Many years later, he remembered these early student days as the beginning of his doubts on the doctrines proposed by the influential Greek philosopher Aristotle, which stated that heavier things fall faster than light ones. He asked one of his professors during a lecture why hailstones of different sizes strike the Earth at the same time, when they all started their fall together. According to Aristotle, the bigger hailstones should have arrived first and the smaller ones last. However, this was not what was observed. The young student was asking awkward questions and the professor was not pleased by such impertinence and told Galileo that hailstones of different sizes originated at different heights, leading the puzzled young man to marvel at the fact that, having all started off in different places, all hailstones hit the ground together! Galileo did not believe this theory: He had begun to see beyond Aristotle.

He was increasingly drawn to mathematics, and slipped into the lectures given by Ostilio Ricci (1540–1603), the mathematician to the

Above *Lady Arithmetic. It was believed that arithmetic, like many other "sciences" were ruled by divine power.*

Medici Court. For Ricci mathematics was no abstract subject, but a practical way to describe the universe, and Galileo believed that mathematics would replace Aristotelian rules. His absence from his own medical lectures was noticed by his tutors, as was his insolent attitude. In the summer of 1583, Galileo asked Ricci to talk to his father and persuade him to let him drop medicine and take up math. Vincenzo, however, was unconvinced and kept his son waiting for a year for his decision. Meanwhile Galileo went to lectures on medicine during the day, and in his spare time was taught by Ricci.

A story exists about the nineteen-year-old Galileo that shows he was beginning to discover the hidden order of nature, and the power mathematics had to describe the physical world. It is said that he was attending services one Sunday in Pisa Cathedral. Above the drone of vespers and the waft of incense, he noticed an oil lamp flickering in the breeze from outside. He saw that it was swaying back and forth in a regular manner, and he measured the time of the swing using his pulse. He rushed back to his cousin's house, cut different lengths of string, and prepared different weights to investigate the properties of pendulums. Did the length of the string matter? Did the amount of weight at the bottom affect the timing of the swing?

Later Galileo claimed, incorrectly as it happens, that a pendulum's swings always take the same amount of time, independently of the amplitude. This was however only true for small amplitude swings. After his death, Galileo's son, also called Vincenzo, sketched a clock based on his father's theories that was however never built. Today you can still stand under the great dome at Pisa Cathedral and watch the lamp, now with electric lights, swing back and forth. It is called "Galileo's lamp." His pendulum clock impressed the university authorities who thought it could find a role in medicine, and for a time it put Galileo back in the good graces of the university tutors, who promptly stole his idea.

Vincenzo had but one hope left for his son to become a doctor. The Duke of Tuscany was offering forty scholarships to needy

Above *Pendulum in motion.*

students. Vincenzo hoped Galileo would be one of them as by then the twenty-one-year-old had invented a significant medical instrument and his father was a noted musician at the Court. Unconvinced, Galileo allowed a request to go forward, but because of his reputation as troublesome, it was refused. For Vincenzo this was something of a crisis, as his dreams of family stability and of a financially secure future evaporated. Galileo left the University of Pisa without a degree and returned home to the wool trade. It must have seemed that he was a great disappointment to the struggling family, but Galileo knew that he would never have become a doctor. Therefore, for the next four years he did what he could to avoid being a drain on the family resources. This involved becoming a private tutor in mathematics to rich families whose male offspring were attempting to pass university examinations.

Despite being far from those who shared his passion, he did not neglect his own development in mathematics. Galileo was very interested in the work of Archimedes, the Greek philosopher, who reputedly ran naked through the streets of Syracuse shouting "Eureka" after realizing that while settling into his bath he displaced a volume of water equal to the volume of his submerged body. Galileo wrote, "Those who read his works realise only too clearly how inferior all other minds are compared to Archimedes." Legend has it that Archimedes constructed a scale and by determining the weight and volume of King Hiero's crown, he proved that that the gold had been mixed with baser metal. Galileo thought he could improve Archimedes' scale: his was small and delicate and had a thin wire as its measuring arm. He called it "the little balance," or "la bilancetta," and when years later it was reproduced in glass, everyone remarked how exact and beautiful it was. It was the most accurate measuring device that had ever been made.

Meanwhile his father was battling in support of new musical tenets. This conflict was to influence young Galileo deeply. His father was fearlessly going against authority and tradition. Indeed, to understand something of Galileo's early upbringing, consider a quote from Vincenzo *(right)*. How often would his son repeat those very same words throughout his own life?

"It appears to me that those who rely simply on the weight of authority to prove any assertion, without searching out the arguments to support it, act absurdly. I wish to question freely and to answer freely without any sort of adulation. That well becomes any who are sincere in the search for truth."
Vincenzo Galileo

Vincenzo was a true revolutionary. He was among the first to explore the emerging key system in his musical compositions that he favored over the restrictive church modes in use at the time. He recognized the superiority of equal-tempered tuning, and compiled a codex illustrating the use of all twenty-four major and minor keys as early as 1584. His essays on acoustics anticipated several of the findings about the nature of sound that would be made by his famous son, and he carried out what was arguably the most important scientific experiment that had been performed until then.

Everyone who had knowledge of music in the sixteenth century had learned that it rested upon the philosophy of Pythagoras. It is said that the Greek philosopher was walking home one day past a blacksmith's shop and noticed that the note given off by hammers of different sizes was different—the larger the hammer the lower the note. This observation led Pythagoras to experiment with strings of animal gut of differing lengths, and to the discovery of "the laws of octaves" that linked what was pleasing to the ear to the mathematics of music. Everyone learning the rules of music knew this, but for Vincenzo the rules did not make sense. He knew that when groups of singers came together to sing in counterpoint the strict octave-based rules of Pythagoras did not seem to work, because they produced dissonant results that sounded inharmonious to the ear.

Galileo chanced upon his father one day as Vincenzo attempted to reproduce Pythagoras' experiment. He joined in, twisting gut, cutting each strand to length, and then stretching the strings over wooden frames and tying them down with weights. Together they tweaked the strings listening to the pitch of the notes they made as they altered the tension on them. They soon realized that Pythagoras was wrong! Pythagoras had said that the pitch varies inversely to the weight attached to the string. However, what Galileo and his father were discovering was that the pitch varied with the inverse square of the tension. It is hard to underestimate the importance of this moment in Galileo's life. He and his father had found a new harmony; a new set of mathematical laws that correlated the note produced by a string to its tension and had done so by experiment. They had not looked up the answer in

Above *A portrait of Pythagoras.*

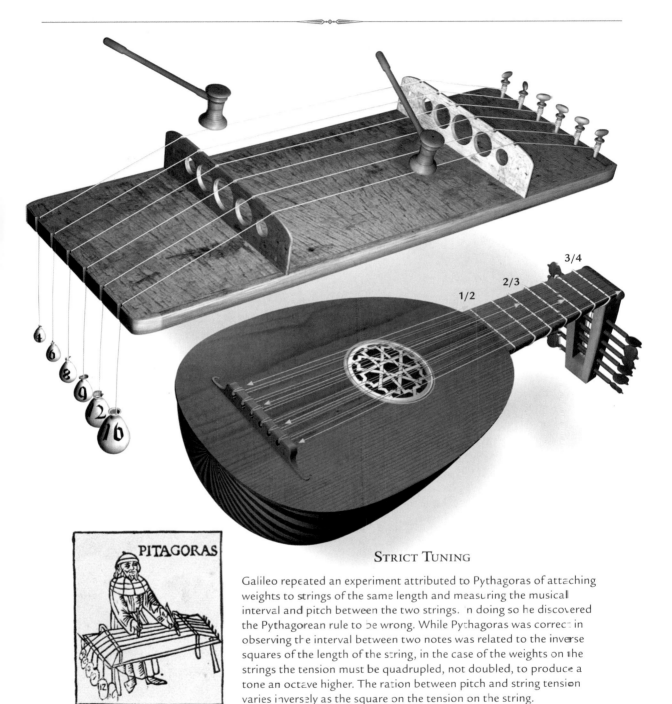

3/4

2/3

1/2

4
6
8
9
2
16

STRICT TUNING

Galileo repeated an experiment attributed to Pythagoras of attaching weights to strings of the same length and measuring the musical interval and pitch between the two strings. In doing so he discovered the Pythagorean rule to be wrong. While Pythagoras was correct in observing the interval between two notes was related to the inverse squares of the length of the string, in the case of the weights on the strings the tension must be quadrupled, not doubled, to produce a tone an octave higher. The ration between pitch and string tension varies inversely as the square on the tension on the string.

PITAGORAS

Above *Traveling musicians.*

either an ancient Greek treatise nor sought the advice of some musical authority. This was the start of modern science: They had carried out an experiment and asked a question of nature itself. It was revolutionary. Vincenzo's actions had unfolded the course of his son's life in experimental physics.

Soon Vincenzo applied these new findings to the playing of his lute, and determined that the best sound could be produced if the frets were of different sizes, getting progressively smaller than the preceding lower fret. He outlined his ideas in *Dialogo della Musica Antica e della Moderna* (1591), in which he attacked the elaborate polyphonic style of the sixteenth century. His discoveries were an important contribution to the efforts of the Florentine Camerata, the local circle of musicians. The musicians of Florence were responsible for the birth of opera, and Vincenzo played a role in its development. Music at the time divided into two different forms: medieval liturgical dramas—holy plays performed publicly at various times in the church calendar—and classical Greek plays that had been revived and were performed with musical interludes. Combining the two forms resulted in opera: non-religious music that combined music and drama. The term takes its name from *opera in musica*, or "work in music."

No doubt Galileo, contemplating his future career, read the closing words of his father's second treatise (*right*).

Galileo's situation changed: he had shown much promise in the mathematical experimentations with his father. All agreed that he could not languish as a minor peripatetic math tutor. It was decided that he would return to university and lobby for the Chair of Mathematics. It was 1587 and the University of Bologna had a vacancy: Bologna was one of the oldest and most respected universities in Italy, and indeed in the whole of Europe. He mounted a campaign to secure the post, a campaign that took him to Rome for the first time. He wanted the patronage of one of the most famous mathematicians of the day, a Jesuit in the Collegio Romano, known as the Euclid of the sixteenth century— Christopher Clavius (1538-1612).

Clavius was German, a Jesuit mathematician, an astronomer, and the main architect of the modern Gregorian calendar that just five years previously had replaced the increasingly off-season Julian

"Now if someone reading this discourse of mine fears to drown traversing with his feeble wings the river become suddenly so wide and rapid, he should follow the example of the great Philoxenus and turn back to traverse it where it takes its origin, until I, myself or others draw the arches over the fundamental I have cast, stretching across its bank a bridge spacious and capable of taking everyone."

Excerpt from Second Treatise
Vincenzo Galilei

almanac. He was among the most respected astronomers in Europe, and his textbooks were used for over fifty years for astronomical education throughout the continent. On Galileo's first visit to Rome, he found to have much in common with Clavius in their mutual passion for mathematics. In his own youth, the Jesuit had suffered the ridicule of math from philosophers and theologians. He had set about changing their perception, and become an acclaimed mathematician. If Galileo could elicit Clavius' support, he would undoubtedly be successful in getting the post at Bologna.

Right *The compass seen as the eye of God.*

When the Vatican initiated the reform towards a historical Gregorian calendar, it also began to take a more active interest in science. Clavius often worked in the Tower of the Winds, a remarkable series of rooms designed by the late mathematician and papal cosmographer Ignazio Danti (1537-1586), whose Chair at Bologna University Galileo now sought. Perched high above the Vatican library, the tower encloses an array of richly decorated frescoes as well as strategically placed holes in the walls so that rays of sunlight can fall upon the meridian marks on the floor at the change of the seasons. Clavius used an open gallery on the perimeter of the tower for astronomical observations. A century later, it would house the Vatican's telescope.

The young, ambitious, but untried contender was ushered into the presence of the great man, a short, stout, and unpretentious person. Clavius' room was adorned with piles of books and charts, many scientific instruments, compasses, quadrants, astrolabes as well as geometrical solids. The objects in the room echoed Galileo's quest of finding the center of gravity in solid objects. His were only ideas; the young mathematician did not yet have the ability to turn them into mathematical theorems. Despite the vast gulf of experience between them, Clavius treated the young man with respect, but seemingly failed afterwards to promote his case for the post at Bologna. They were to meet again many years later, Galileo holding a telescope in his hand.

The Bologna post went to Antonio Magini (1555–1617), a Paduan twelve years older, who was in every way better qualified than Galileo, having already written several mathematical books. He never forgot the disappointment of losing to Magini, and when competing against him again several years later, Galileo was far better prepared.

Vincenzo died three years later. He was laid to rest in Santa Croce Church, the burial place for the great and the good in Florence, where Michelangelo and Machiavelli were buried, and where the empty sarcophagus of Dante stood. Fate would have it that Galileo himself would be buried next to Vincenzo, but not until almost a century after his death. From the day of his father's passing, the lute never left Galileo's side.

Galileo decided to follow his father's example and try his luck at the court of the Grand Dukes of Tuscany. After all, his father had been an influential figure at Court and Grand Duke Francesco de' Medici was interested in science. However, suddenly Francesco I and his wife died in questionable circumstances, and the new

Above *A detail of the Santa Croce Church and Monastery where Vincenzo Galilei was buried.*

Above *An engraving showing Galileo's pendulum clock.*

Left *The brass instrument represents the five Platonic solids.*

Duke was less taken by Galileo's interests. Fortunately, another opportunity arose: The Chair of Mathematics at the University of Pisa was being revived after years of vacancy. This time Galileo would not be as ill prepared as he had been when contending for the post at Bologna.

To make a successful bid he needed the patronage of the rich and influential, and he befriended Marquis Guidobaldo del Monte (1545–1607) who was both a nobleman of repute and an accomplished mathematician. He used the questions previously asked of Clavius as common ground to establish a relationship with del Monte. Galileo knew that the Marquis' family would prove useful. On July 16, 1588, he wrote to him *(right)*.

The Florentine intelligentsia gathered at the Academy, and had done so for fifty years to consider the works of Boccaccio and Dante. Members of the Academy had become interested in the question of the size and location of Dante's Inferno. Galileo got to hear of this and pursued the chance of appearing before Florence's richest and most influential men. Standing on the podium in the Medici Palace in the Via Larga, Galileo began by underscoring the difficulties of measuring and locating Dante's Inferno. Taking his clues from the great poem in the *Divine Comedy*, he explained that Dante's hell was cone-shaped, with its vortex being the home of Lucifer whose lower parts were locked in ice and whose belly button marked the very center of the Earth. Galileo added, "There

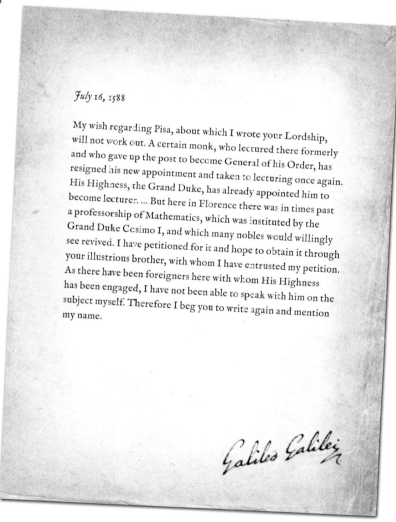

July 16, 1588

My wish regarding Pisa, about which I wrote your Lordship, will not work out. A certain monk, who lectured there formerly and who gave up the post to become General of his Order, has resigned his new appointment and taken to lecturing once again. His Highness, the Grand Duke, has already appointed him to become lecturer. ... But here in Florence there was in times past a professorship of Mathematics, which was instituted by the Grand Duke Cosimo I, and which many nobles would willingly see revived. I have petitioned for it and hope to obtain it through your illustrious brother, with whom I have entrusted my petition. As there have been foreigners here with whom His Highness has been engaged, I have not been able to speak with him on the subject myself. Therefore I beg you to write again and mention my name.

Galileo Galilei

Above *A painting by Sandro Botticelli of Dante's Inferno, whose size and location Galileo attempted to find.*

is a relation between the size of Dante and the size of the giant Nimrod, in the pit of hell, and in turn between Nimrod and the arm of Lucifer. Therefore if we know Dante's size, and Nimrod's size, we can deduce the size of Lucifer." Working through this logic Galileo concluded that Lucifer was 1,935 human arm-lengths tall. His lecture before the Academy of Florence was a success, but he still did not have a job. Over the years, he had applied for posts at the universities of Siena, Padua, Pisa, and Florence, and was turned down by all of them. Perhaps it was time to leave Italy, and seek an academic job abroad, perhaps in the Middle East. He despaired. Would he ever get the chance to pursue his ideas?

All of a sudden his fortunes turned. The Chair of Mathematics became available once more at the University of Pisa and he swiftly used the influence of Marquis del Monte and his other famous patrons. This time he got the post. There were caveats however: It was only for three years and the salary was a paltry sixty crowns per year and his family would eat up most of it. Moreover, Pisa had a far lesser reputation than Bologna. Most of its students enrolled in law and he found that those who came to his classes were largely unenthusiastic about the subject of mathematics. He took rooms in a low building on the north side of the river Arno. Galileo could not help wonder whether Pisa was to be a springboard for his ideas or the backwater in which he would labor without notice. He annoyed the university authorities with his arguments and his refusal to wear academic dress.

Once again, Galileo thought about the hailstones and the motion of falling bodies and Aristotle's statement that heavier objects fell faster than lighter ones. One day, when lost in contemplation, he looked up and saw the leaning Tower of Pisa. The sight kindled his imagination.

CHAPTER TWO

THE HIDDEN LAWS OF NATURE

In the Padua years, Galileo conducted studies and experiments in mechanics, he built the thermoscope, and invented and built the military compass. In 1594 he also patented a water-lifting machine. Despite these small successes, however, Galileo struggled to find a big invention that would assuage his more personal family problems.

At the end of the final moonwalk of the Apollo 15 mission in July 1971, Commander David Scott stood in front of the remote-controlled television camera to perform a live demonstration for viewers back on Earth. With the lunar Apennine Mountains gently rising behind him he said, "One of the reasons we got here today was because of a gentleman named Galileo." Scott held a geologic hammer in one hand and a feather in the other. He released them at the same time, and, being in a vacuum with no air resistance, the feather and the hammer struck the lunar soil at exactly the same speed. "How about that," said Scott, "Galileo was correct." It was a space-age re-enactment of one of Galileo's most famous experiments.

Nearly 400 years earlier, it is said, Galileo climbed to the top of the Tower of Pisa carrying balls of various compositions and weights. It must have taken several trips up the winding staircase to transport the balls of lead, ebony, porphyry, and copper to one of the open verandas near the top of the tower. What better place was there to conduct an experiment to see if Aristotle was correct in believing that heavier objects fell faster than lighter ones? Aristotle had written that, "...a one hundred pound ball falling from a height of one hundred cubits reaches the ground before another of one pound has descended a distance of one cubit." Galileo did not believe it. In fact there was little of Aristotle that he did believe. So he prepared to test the idea from the Tower of Pisa that was almost a hundred cubits high.

Below *Astronomer with astrolabe.*

Above *A graphic representation of Commander David Scott of the Apollo 15 Mission, reenacting an experiment Galileo performed centuries before to prove the force of gravity at work on all objects, whether light or heavy.*

The story goes that he publicized his experiment widely and that quite a crowd of skeptical professors and enthusiastic students gathered below to watch. As he emerged onto the balcony he was greeted by a roar of approval, though few in the crowd doubted that Aristotle would not be vindicated. Many had arrived wondering what the fuss was about, as they had never heard of the young exhibitionist parading before them.

Suddenly Galileo released two balls of very different weights. They plummeted swiftly and struck the ground together. The crowd gasped. Galileo smiled. Aristotle was wrong.

But did Galileo's most famous experiment really happen, or is it another legend like the story of the pendulum? It is strange that Galileo did not write about the incident at the time. The tale comes from his first biographer Vincenzo Viviani (1622–1703), who began transcribing Galileo's experiences when he was already old.

If he did not actually conduct that experiment from the Tower of Pisa, had he done so, it would have been entirely in keeping with his exhibitionist and rebellious character. Throughout his life, Galileo had little regard for his superiors, and one of his perennial targets was Aristotle, the ultimate authority for university philosophy faculties at the time. Galileo's personal style was confrontational, witty, ironic, and often sarcastic. His intellectual approach, as the Tower of Pisa story suggests, was to construct his theories with an ultimate appeal to observations. Much later Galileo was to write, "I say, that the balls reach the ground at the same time. In doing the experiment, you find that when the heavier mass touches the ground, the lighter is two fingers away."

At the University of Pisa, Galileo became interested in motion. He experimented by dropping objects from a certain height, and rolling balls down a gently sloping inclined plane determining their positions after equal time intervals. Just as with the experiment he had conducted with his father with the strings, he saw that there was a hidden mathematical order beneath reality.

He chronicled his discoveries about motion in his book, *De Motu*, which literally means "On Motion." In retrospect, his first significant piece of writing is an immature work, punctuated by a few flashes of brilliance that however floundered in the light of a true theory of freely falling bodies. He never published it.

"He upheld the dignity of his professorial chair with so great fame and reputation, before judges well disposed, and then since many of his rivals, stirred with envy, were aroused against him."
Viviani

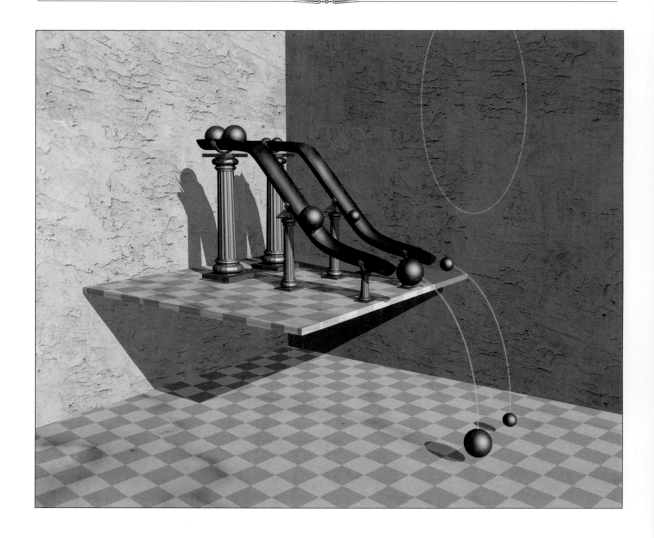

ROLLING INCLINES

Galileo conducted an experiment in which he
rolled balls of different weights down an incline.
His measurements showed that each body
increased its speed at the same rate. He also
showed that the trajectory of the final fall onto
the floor was elliptical.

With his father's death, Galileo, reluctantly it seems, became head of the household, which meant that everyone was dependent on him for money.

The truth was that during his three years at Pisa, he had made little real scientific progress and probably many enemies. The only friend he had among the professors was Jacopo Mazzoni (1548–1598), a man of enlightened views, who had been appointed to the Chair of Philosophy the year before Galileo's own appointment took place. With this exception, the whole body of professors was hostile to him and they, as well as the heads of the university, were staunch Aristotelians. Although he remained unimpressed and even in contempt of the establishment and displayed a natural tendency to rebel, he was not yet a revolutionary. There were rules he could break, and be seen to break. He refused to wear academic gowns and went about in tatty, disheveled clothes.

Galileo increased his unpopularity when he was critical of a hydraulic pumping machine devised by Don Giovanni de' Medici (1498–1526), then governor of the port of Leghorn (Livorno). Galileo said it wouldn't work. When the device failed completely, many in the Medici entourage resolved to do all in their power to ruin the young professor. Unsurprisingly, his contract with the University of Pisa was not renewed, but it seems he did not mind as he already had something else within his sights. In the summer of 1592, he began campaigning for the Chair of Mathematics at the University of Padua.

Padua was in a much higher league than Pisa. It was the oldest university in Italy after Bologna, and its Chair of Mathematics was one of the most prestigious posts in Europe. It had been vacant for years after the death of the renowned mathematician Giuseppe Moletti (1531–1588).

Galileo, however, was an outsider in Padua, and at a disadvantage in coming from Tuscany

Below *Portrait of Don Giovanni de' Medici.*

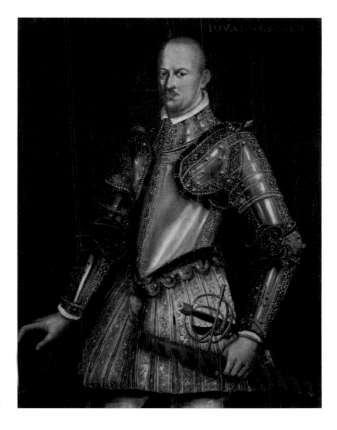

Above *Portrait of Tycho Brahe.*

when it came to applying for a post under the influence of the Serene Republic of Venice. To make matters worse, Antonio Magini, to whom Galileo had lost in competing for the Chair at Bologna years before, now also wanted the Chair at Padua. Galileo decided that he needed a charm offensive. Topping the list of influential people Galileo needed to charm was Gianvincenzo Pinelli (1535-1601), who not only surrounded himself with a group of writers and intellectuals, but who also possessed a fabulous library of some 80,000 tomes. Soon Pinelli was on Galileo's side, advising him on how to tackle the Venetian Senate. With the additional help of the Tuscan ambassador to Venice, Galileo presented an imposing figure and won the post. Magini went back to Bologna and bided his time to exact revenge. So it was that Galileo moved to the city that Shakespeare had called, "fair Padua, nursery of the arts."

He was determined to make a grand entrance. So it was that on December 7, 1592 he stepped up to the lectern in the Great Hall of the university to give his inaugural lecture in front of his robed peers. He had been preparing the lecture, to be presented in Latin, for weeks, committing the most important passages, as well as the jokes, to memory. It was a resounding success, and in the following months his reputation spread and the most famous astronomer of the day, Tycho Brahe (1546-1601), working in Denmark, said that a new star had appeared in the firmament.

He initially stayed at Pinelli's house where he met the leading intellectuals of the university, who questioned almost everything in established truths of the time. He met rebels such as himself, and what happened to them did not go unnoticed to Galileo.

One was Tommaso Campanella (1568-1639) who wrote the *Philosophia Sensibus Demonstrata* (Philosophy demonstrated by the senses), in 1592. Because of his unorthodox views, Campanella spent twenty-seven years in prison. Another person Galileo may have met at that time was Giordano Bruno (1548-1600), who later influenced Galileo's thinking considerably. Bruno was the first who since the triumph of Christianity preached a return to the independence of Greek thinkers. Soon he was seized by the Venetian Inquisition and handed over to Rome on charges of heresy.

Years later, Galileo would experience the Inquisition at first hand.

Right *Portrait of Nicholas Copernicus.*

In the fourth century, as the Holy Roman Church was forming its doctrines, it issued fifteen edicts against heresy—the *Imperatoris Theodosiani Codex*—and set up the Inquisition in the thirteenth century. Anyone could make a denunciation against anyone and the accused knew neither the crime nor the accuser. It became a way of life and death in medieval Europe. In 1542, reeling from the works of Luther, Pope Paul III set up the Roman Inquisition based on the ruthless Spanish ecclesiastical tribunal that had been established by the Castilian monarchs almost a century before, but events were already overtaking him. Just a few months after the Holy See's tribunal for trials of heresy was founded, Nicolaus Copernicus (1473-1543), in what is arguably the most controversial book of all time, claimed counter to the teachings of the Church that stated that the sun went around the Earth. His book, *De Revolutionibus Orbium Coelestium* (On the revolutions of celestial spheres), was ironically dedicated to the same pope. Those who had faced the Inquisition and miraculously survived to tell the tale had the following to say to others about to go to trial: "All witnesses and officials are bound by oath to eternal silence. Answer a thousand questions with a thousand identical answers. Fall just once into a trap set a thousand times." Galileo knew that Giordano Bruno fell, deliberately. The crime he was accused of was of believing heretical things about the universe, deeming it bigger than scripture said and that it might contain other worlds and other peoples.

In September 1593, the Inquisition demanded Bruno's extradition, and in the following year, he was held prisoner in the dungeons near the Vatican. Only in 1599 did the Roman Inquisition start his trial. A decree of February 4, 1599 ordered him to recant his eight heresies. He did so, but shortly afterwards suffered a nervous breakdown. He challenged the authority of the Holy Office and appealed directly to the Pope. But Pope Clement VIII wanted rid of the priest who held uncomfortable heretical ideas. On January 20, 1600 he signed the document that handed him over to the secular arm of the church. Bruno was excommunicated once more and his soul decreed worthless. Cardinal Bellarmine (1542-1621), who was canonized in 1930 as both a saint and Doctor of the Church, and who was later to cause

Galileo so much trouble, personally signed the death warrant of Giordano Bruno. The priest was dragged to the Campo dei Fiori and burned at the stake. What remained of his body was gathered, smashed to powder, and cast to the wind.

Galileo was also well aware of ill-fated Antonio de Dominis (1566–1624), who had been born in Croatia. In his youth he showed enthusiasm for learning and was sent to the Illyrian Jesuit College in Loreto. Later, he would also teach mathematics at Padua. Today, de Dominis is not remembered as much for his scientific studies as by his folly. Traveling to England he renounced his Catholic faith in favor of Protestantism, but lacked the sense to stay out of the reach

Above *Heretics being burned at the stake, the fate of those proven to have disagreed with the Church by the trials of the Holy Inquisition.*

Above *Portrait of Cardinal Bellarmine.*

of Rome. Wishing to reconcile with the Holy See, he left England and returned to Rome. His subsequent attacks against the English Church were as violent as his previous inveighing against Rome. He said that he had deliberately lied in all that he had said against Rome. Pope Gregory XV believed him and even gave him a pension, but there were others who did not. When the Pope died in 1623, Dominis came into conflict again with the Inquisition. He was finally imprisoned in 1624 and accused of having returned to his heretical beliefs. He was given three months to prepare his defense, but became seriously ill and died on September 8, 1624. His case was continued after his death: Dominis was declared a relapsed heretic, and his body was exhumed and burned together with his works on December 21 at the Campo dei Fiori, on the same spot where Giordano Bruno had been burned in 1600. While Bruno was declared a martyr to science de Dominis was not.

Galileo had the opportunity to observe the Church's retribution against scientists very well during his life.

Often at Pinelli's intellectual soirées Galileo would play the lute and impress the guests. This was a place to see and be seen, and to make useful contacts as Galileo contemplated various schemes to make money to support his family. It was here that he met Giacomo Contarini (1536–1595), who was in charge of shipbuilding in Venice. Shipbuilding was the great industry of Venice, and the speed at which the Arsenale—the inner harbor where Venetian ships were fitted out—could produce ships was legendary. Hulls of galleys entered at one end and within a few hours left at the other end, fully equipped and manned. The Venice Arsenale had been a place of practical invention and innovation for centuries. Galileo had always been interested in mechanical things, and at the Arsenale he learned a great deal. His private lecture notes and other writings of this period show a concern for problems in fortification, mechanical devices, and other aspects of technology. Contarini had

Above *Saint Mark's Basin with a view of the Molo, the Piazzetta and the Doge's Palace in Venice.*

Above *Portrait of Ptolemy, the Roman mathematician, geographer, astronomer, and astrologer, whose treatises were to influence European Renaissance scientific thought. This image is from a fifteenth century Greek manuscript of Ptolemy's Geography, and he is shown using an astrolabe.*

some questions about the design of the ships they were producing. Where was the best position for the oars in relation to the hull of the ship? How long should they be? Galileo submitted a report in which he correctly treated the oar as a lever and the water as the fulcrum.

Galileo pondered these technical questions, and attempted to put his own personal problems to the back of his mind. The biggest one was money. He was head of the Galilei family and responsible for the payment of the dowry of his sister Virginia to Benedetto Landucci. He had missed several payments and he received an urgent letter from his mother *(below right)*. In order to pay Benedetto Galileo had to go deeper into debt. He had to take on many private students to keep himself afloat. Twenty-eight-year-old Galileo moved into a small house, and contemplated other ways of increasing his income.

At the time, there was no way to quantify heat. In Aristotelian theory, heat and cold were fundamental qualities. Like dry and wet, heat and cold were qualities combined with *prima materia* to make up the elements—earth, water, air, and fire. Thus earth was dry and cold, fire dry and hot, and so on. No distinction was made between the concept of heat and the concept of temperature. Measuring heat became a puzzle and a challenge in the circle of practical and learned men to which Galileo belonged. The first solution was a device called a thermoscope, which was based on the idea of the expansion of air as its heat increased, and vice versa. The first versions were little more than toys.

It was reported that Galileo took a small glass flask, about as large as a small hen's egg, with a neck about two spans long (perhaps 16 inches) and as fine as a wheat straw, and warmed the flask well in his hands, then turned its mouth upside down into the vessel placed underneath, in which there was a little water. When he removed the heat of his hands from the flask, the water at once began to rise in the neck, and mounted to more than a span above the level of the water in the vessel. Over the next few years, this

Above *The Arabic astronomer Massahalla in a drawing by Albrecht Dürer, from Stebius' Scientia, Nuremberg, 1504, who wrote a treatise on the astrolabe that was used as a scientific source in Renaissance Europe.*

"You should know that Benedetto wants his money now and is threatening to have you forcibly arrested immediately when you arrive here. He is just the man to do it, so I warn you: it would grieve me much if anything of the kind were to happen."
Galileo's Mother

early thermoscope was improved and Galileo added a numerical scale making it the first air thermometer, but this invention did not earn him much money.

He hoped to earn more from an irrigation mechanism he constructed. The Venetian Senate awarded him a patent for a water lifting machine raising water by means of one horse *(below left)*. Despite the patent, he never made money from this invention. However, he was learning to apply practical skills to mechanical devices, something that would prove very useful when it came to constructing a telescope.

Padua offered a far more congenial atmosphere for Galileo's talents and lifestyle than the intellectual backwater he had found in Pisa. In the nearby city of Venice, he found recreation and many friends. Galileo's favorite debating partner among these was Gianfrancesco Sagredo (1571–1620), a wealthy nobleman with an eccentric manner he appreciated. He was a Venetian aristocrat nine years younger than Galileo. He lived in a palace along the Grand Canal and he also had a country estate where he threw wild parties. He once wrote, "I am a Venetian gentleman. I have never called myself one of the literati, but hold dear the protection of the literati. Never have I intended to take advantage of my fortune. My studies always concern those things which, as a Christian, I owe to God, as a citizen, I owe to my country, as a nobleman, I owe to my house, as a man of society, I owe to my friends, and as a gallant man and philosopher, I owe to myself. My palace in Venice has often been compared to Noah's Ark, partly because of its shape, partly because I

"That by the authority of this Council is granted to Mr. Galileo Galilei that for the space of the next twenty years others than him or his agents are not allowed in the city or any place in our state to make, have made, or, if made elsewhere, to use the device invented by him for raising water and irrigating fields, by which with the motion of only one horse twenty buckets of water that are contained in it run out continuously."
The Venetian Senate

Above *The University of Padua.*

keep inside all manner of beasts. As a bachelor, I spend my time in conversation. If occasionally I speculate on science, don't believe, dear sir that I presume to be equal to the professors. I investigate freely the truth of those propositions which give me pleasure."

During his time in Padua Galileo made some of his most important discoveries in mechanics and astronomy. From careful observations, he formulated a law that states that the vertical distance covered by an object in free fall or along an inclined plane is proportional to the square of the time of the fall.

His time at Padua was perhaps the best of his life. Although he was continually afflicted by financial problems, he was well established and respected, free to carry out his research and his personal indulgences. He used its taverns, prayed in its churches, walked its public spaces, gazed at Giotto's masterpieces, and debated with his students on the university's marble steps.

Above *The world principles of the world's systems according to Brahe, Copernicus, and Ptolemy represented here in the form of clocks.*

Galileo seemed to enjoy this time even carelessly. In the summer of 1593 along with three friends he set off for a trip to the countryside. They went to Verona and walked to the small village of Costozza, where they spent the night in the villa of a lawyer friend who was also a patron of the arts. After a night of heavy drinking he lay down naked on the grass next to a cave from which a cold draft was blowing. He caught a chill and had to be taken back to Padua in a litter and confined to bed with the malady as well as arthritis. Friends said he was lucky to survive.

The following year Galileo became interested in astronomy. In 1595 he was sent a book by Johannes Kepler (1571–1630), a young scientist with great potential. The book was called *Mysterium Cosmographicum* (The Cosmographic Mystery) and it contained Kepler's ideas about the sun rather than the Earth being the center of the universe. Although he found it a rather confused book, Galileo was impressed and wrote Kepler *(below right)*. Kepler wrote back urging him to publish, "..the truth must be all-convincing," he wrote. "Publish in Germany" if he were not allowed to do so in Italy, he said. This, however, Galileo did not do. Although he discussed the Copernican theory privately, and succeeded in convincing many of his friends of its superiority to the Earth-centered theory, he continued to teach that system for several more years.

In 1597 Galileo began collaborating with the toolmaker Marcantonio Mazzoleni (?–1632), a relationship that was to last ten years. Galileo's old friend, the Marquis del Monte, had wondered if it would be possible to develop a simple compass that could be used to gauge the distance and height of a target as well as to measure the angle of elevation of a cannon's barrel. Galileo, with Mazzoleni's help, produced a lightweight, brass instrument that they began to produce for sale. Galileo was delighted. He had, at last, developed a money-spinning device, and he could not only charge for it but also for lessons on how to use it, 120 lire a time. In 1599 Mazzoleni moved into Galileo's house with his wife, who became the cook, and his child.

Galileo had learned from his previous setbacks, and when his first tenure at the University of Padua expired in 1599 he was sufficiently well regarded for tenure to be extended for another seven years. He was in his mid-thirties, successful, and living in a comfortable house.

However, events that had occurred before he was born and taken place far from where he had lived were to change his comfortable life in the next few years. The sixteenth century had not been a good period for the Catholic Church. It had been the time of Luther and Henry VIII. In Rome, 1600 was to be a year of renewal and festivals. In Jesuit circles Cardinal Bellarmine gave well-attended lectures.

Above *A portrait of Johannes Kepler, the influential German astronomer.*

It is really pitiful that so few seek truth, but this is not the place to mourn over the miseries of our times. I shall read your book with special pleasure, because I have been an adherent of the Copernican system for many years. It explains to me the causes of many appearances of nature which are quite unintelligible within the commonly accepted hypothesis.

I have collected many arguments to refute the Aristotelian theory, ... but I do not publish them, for fear of sharing the fate of our master Copernicus. Although he has earned immortal fame with some, with many others (so great is the number of fools) he has become an object of ridicule and scorn. If there were more people like you, I would publish my speculations. This not being the case, I refrain.

Galileo Galilei

The Florentine Pope, Clement VIII, looked forward to the new century. Henry IV of France had renounced Protestantism. The influence of Spain was declining. The heretic Queen Elizabeth of England was isolated. The last thing Rome wanted was more trouble.

Financial problems persisted for Galileo. His extravagant younger brother Michelangelo expected him to support his efforts to become a professional musician. In the summer of 1600, Michelangelo was appointed court musician to a noble Polish family. The problem was that his brother needed a collection of impressive and expensive musical instruments to take with him.

It had been the family's intention that his sister Livia, the only unmarried daughter at the time of Vincenzo's death, was to take the veil. Livia, however, wanted to leave the convent, and as she was nearing thirty Galileo knew that a marriage would not be easily arranged and a new dowry expensive for him. He wrote to his mother, "It is impossible for me to consent to this arrangement just at present. Perhaps when Michelangelo got settled in his new post and could contribute a portion of his new salary to a dowry, Livia's marriage might be possible since she is determined to come out and partake of the miseries of the world. Tell her there have been queens and great ladies who have not married till they were old enough to be her mother."

Eventually Galileo found a husband for his sister and Michelangelo promised to send some money but he never did (*left*). The university authorities refused his request for a pay increase. A few years later Michelangelo lost his job in Poland and came back to Italy to live from Galileo's earnings. He managed to obtain another job for Michelangelo as court musician to the Duke of Bavaria and, for a time, thought finances would stabilize with his brother contributing to Livia's dowry. Then he heard of Michelangelo's extravagance.

"You complain of me having spent such a large sum on one

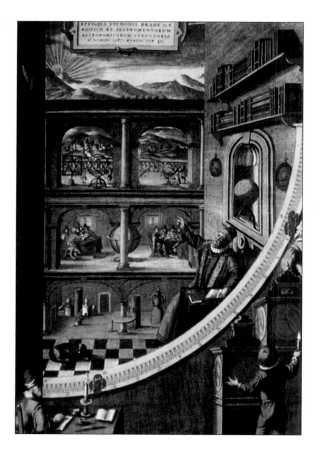

Above *The mural in Tycho Brahe's Uraniborg Observatory, in Denmark, built by the scientist in 1580. Because of the dominance of the Vatican on : all matters, it would have been inconceivable for Galileo to be allowed to build an observatory in Tuscany.*

"If I had imagined that things were going to turn out in this manner, I would not have given the child in marriage, or else I would have given her only such a dowry as I was able to pay myself. I seem destined to bear every burden alone."
Galileo

Above *Allegory of Lady Arithmetic.*

feast," Michelangelo wrote, "the sum was large, yes, but it was my wedding. There were more than eighty persons present, among whom were many gentlemen of importance. Present were no less than four ambassadors! If I had not followed the custom of this country I would have been put to shame. You cannot accuse me of ever having spent such sums of money simply for my own gratification; never, indeed, have I thrown money away on anything, but, on the contrary, have often denied myself what I wanted, in order to save."

Galileo took a mistress, Marina Gamba, described by one of his biographers as "hot-tempered, strapping, lusty, and probably illiterate." Galileo and Marina had three children: two daughters, Virginia and Livia, and a son, Vincenzo. In none of the three baptismal records is Galileo named as the father. In the case of Virginia, she was described as "daughter by fornication of Marina of Venice," with no mention of the father; on Livia's baptismal record the name of the father was left blank; and on Vincenzo's baptismal record it was written "father uncertain." In later life, as tragedy loomed, Galileo found great comfort in the company of his elder daughter, Virginia. Galileo and Marina's domestic situation was, apparently, a happy one, except when Galileo's mother, Giulia, visited. She recruited Galileo's servants to spy on the couple and report back to her.

In September 1604 a remarkable event took place. A new star appeared in the sky, in the constellation of Ophiuchus—the snake holder—located around the celestial equator. A new star was important. One had appeared in 1572 and had been studied by Tycho Brahe. Did it signal that things could change in the immutable heavens? According to Aristotle change was only allowed in the region below the moon, so Galileo wondered if this new star proved once again that Aristotle was wrong. Soon it was brighter than any other star in the night sky and any of the planets as well. Johannes Kepler first saw it on October 17, and it was subsequently named after him due to the fact that he had chronicled its appearance in a book titled *De Stella Nova in Pede Serpentarii* (On the new star in Ophiuchus's foot).

Galileo wrote; "At one point, it would contrast its rays with a sudden and faint extinction, paling before Mars's reddish glow.

Right *Chandra X-ray observatory image of a supernova such as that observed in Padua in 1604.*

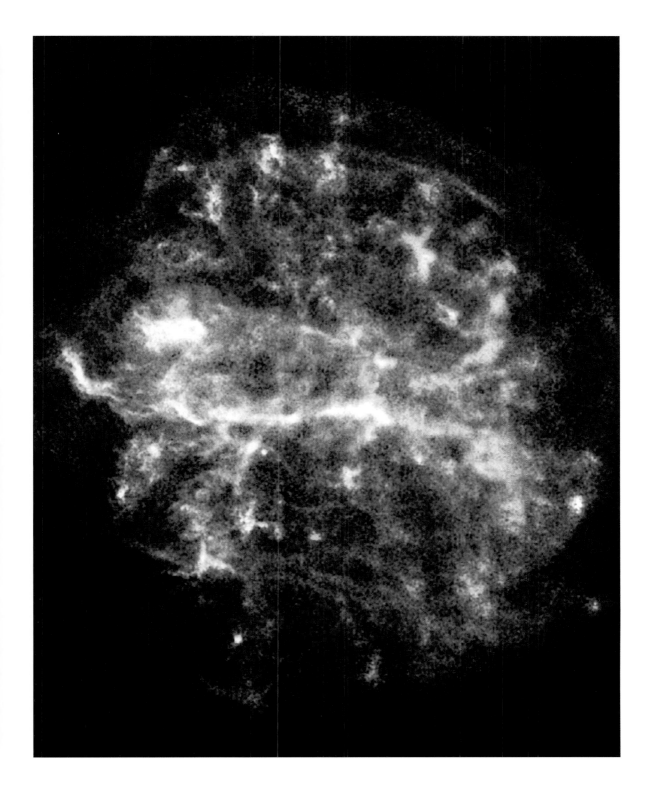

Then straightaway, it would shed even more fulsome rays as if coming back to life, holding forth its own splendor with Jupiter's brilliance. Anyone might believe with good reason that this new light was an offspring of Jupiter and Mars, most especially because it appeared to be born both in the same general direction and at the same time as a predicted planetary conjunction."

To satisfy public interest in the new star Galileo gave three crowded public lectures. He said, "On the 10th October 1604 a certain strange light was first observed in the heavens. At first it was quite small, but soon it was visible even by daylight, surpassing in brightness all fixed and wandering stars with the exception of Venus. It was red as well as sparkling. It gave off waves of light, which seem both to kill and set aflame, more than any of fixed stars and the dog star itself [Sirius]. It had the splendid brilliance of Jupiter and the redness of Mars which is like fire. The contractive quality of these terrible rays announced destruction, as if from the boiling redness of Mars, whilst the expansive quality of these rays gave forth Jupiter's bright lightning. It came forth as the due fruit generated by the intercourse of Mars and Jupiter. In almost the same position as the conjunction had been predicted the new star appeared to be born. The star was located where previously no conspicuous star had been observed."

An academic called Cremonini (1550–1631) asked him, "If we abandon Aristotle, who will be our guide in philosophy?" Galileo replied, "Only the blind need a guide. Those with eyes and those with a mind must use these faculties to discern for themselves."

Soon his mind was back on more earthly matters. Benedetto Landucci had started legal proceedings against Galileo as his dowry had not been paid for two years and Galileo risked being arrested. He not only did not have the money, but was already in debt. This time there seemed no way out. Then the Grand Duchess Christine of Tuscany requested that Galileo instruct her son,

Left *The universe according to Ptolemy. One of the most influential Greek astronomers of his time (c. 165 BCE), Ptolemy propounded the geocentric theory in a form that prevailed for 1400 years.*

Above *Woodcut representing a Christian philosopher holding an astrolabe.*

Prince Cosimo, a bright twelve-year-old, and Galileo was saved once again.

Pope Clement died in March 1605, and with his successor, Pope Paul V, the battle of words between Rome and Venice intensified. In the 1590s and the early 1600s the level of religious and political conflict between Venice and Rome had been rising. Venice claimed to control navigation in the Adriatic, while Rome, backed by the Habsburgs of Spain and Austria, claimed freedom of navigation there. Venice had friendly contacts with non-Catholic states, a fact that annoyed Rome. In 1604 Venice forbade the construction of any new churches or shrines without permission from the state, and in 1605 forbade any further transfers of real property to ecclesiastical institutions without permission from the state. In the summer and autumn of 1605 Venetian authorities arrested two delinquent clerics. Rome had had enough. In January 1606, the papal nuncio delivered a brief demanding the unconditional submission of the Venetians. They refused, and the new pope excommunicated the Doge of Venice, the Venetian Senate, and all Venetian officials, among them Galileo's friend Paolo Sarpi (1552–1623). Sarpi would play a crucial role in introducing Galileo to the telescope years later. He had become notorious as the defender of Venice against the papacy during the Venetian Interdict of 1606. Before this Sarpi had been an obscure ecclesiastic, but after 1606 had become known throughout Europe. He was also highly knowledgeable of the most advanced scientific and philosophical ideas of the time. Sarpi was recruited by the Venetian government to act as its adviser and publicist, and wrote many pamphlets in its defense. After much hard negotiation, in which Sarpi was closely involved, the Interdict was lifted in April 1607. Despite this success, Sarpi was personally excommunicated in early 1607 and, targeted for assassination, was almost killed in October of that year

Below *Bust of Pope Paul V.*

CCXLVIII

when he was set upon by five assailants while walking home. They stabbed him fifteen times and left him to die. Much to the pope's dismay, Sarpi survived.

With such goings on Galileo had to reassess his personal situation. He was well aware that his best interests might not be served if he was too closely associated with rebels and heretics. Perhaps his long-term future did not lie in Venice. He looked for pretexts to contact the Grand Duke in Florence. Gianfrancesco Sagredo, his aristocratic Venetian friend and supporter, departed for Syria and Galileo became increasingly anxious and frustrated, depressed at the need to earn more money with private tuition. "I am always at the service of this or that person. I have to consume many hours of the day—often the best ones—in the service of others." His research had reached a standstill. All too soon his life would change with the arrival of the telescope.

CHAPTER THREE

THE PERSPICULUM

In 1609, a peculiar convergence took over central Europe whereby different spectacle-makers developed magnifying lenses that allowed to observe what was far at close range. However, it was Galileo who perfected the telescope, though others were working on similar instruments before him. With the telescope, Galileo launched scientific observation of the heavens.

The telescope became central to Galileo's life, catapulting him from an obscure, recalcitrant professor into the dominant intellectual force of his time on a collision course with the Church. When he first found out about the instrument, Galileo did not realize its potential to change the world. Like others who had also seen it, he could easily have missed the opportunity.

The precise origins of the telescope are somewhat murky and will probably never be known. Lenses had been produced for centuries, and there are references dating to the thirteenth century as to their capacity to make things far off seem as if they were close by. When Galileo presented his telescope there were some who claimed to have invented it many years previously. Recently a note written in 1608 has been found that suggests a telescope of sorts was already in use. It reads: *Discoverie of the Most Secret and Subtile Practises of the Jesuites* by Johannes Cambilhom (1608) and is accompanied by the following intriguing description: The Society of Jesus' "bawdy adventures with innocent girls, its vast stores of buried treasure, its arsenals of weaponry," along with a "looking glasse of astrology," a "celestial or rather devilish glasse" supposedly used by the Catholic French king "to see playnly what-soever his Maiestie desirded to know."

The Milanese courtier Girolamo Sirtori (?–1631) in his 1618 book *Telescopium, Sive Ars Perficiendi Novum Illud Galilaei Visorium Instrumentum Ad Sydera*, said that he had gone to Gerona, in Spain, and met the real "first inventor" of the telescope, a man called

Below *An astrologer at work with his books and the stars above him. In fact the medieval astrologer rarely studied the stars directly, but took all his data from pre-calculated tables.*

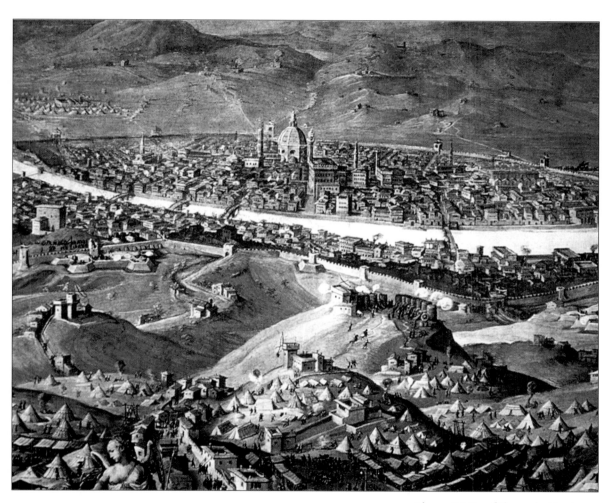

Above *Painting of Florence by Giorgio Vasari, at the time Galileo lived in the city.*

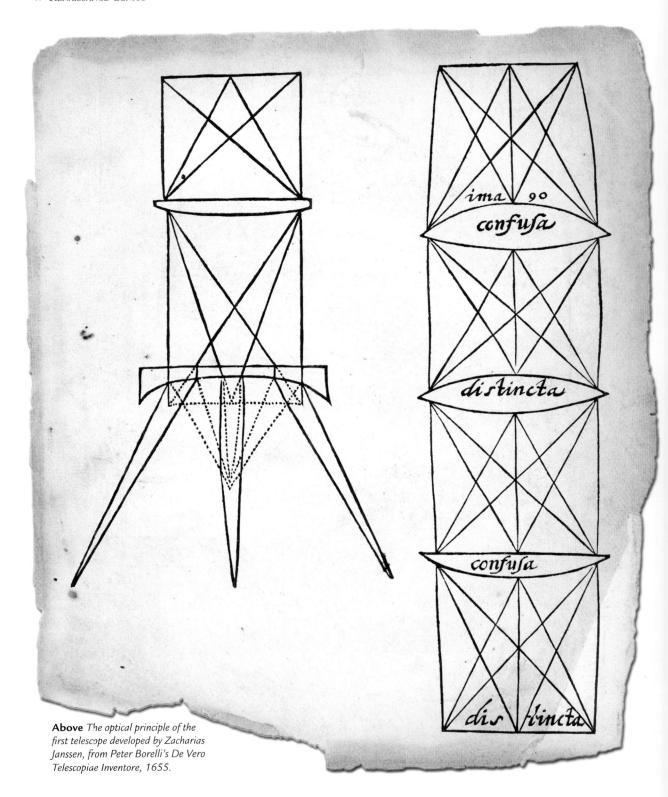

ima 90
confusa

distincta

confusa

dis tincta

Above *The optical principle of the first telescope developed by Zacharias Janssen, from Peter Borelli's De Vero Telescopiae Inventore, 1655.*

Roget of Burgundy. Historians have long thought this too marginal to be significant. However, Sirtori mentions in his book three nobles who died leaving a telescope among their possessions: On April 10, 1593 Don Pedro de Corolona died leaving "A long spyglass decorated with brass." In 1608, Catalonian merchant Jamie Galvany passed away leaving "An eyeglass for long sight." and in 1613, the Marseille-born merchant Honorato Graner left what was described as "una ullera de llauna per mirar de lluny"or "a metal eyeglass for seeing far away."

Despite the lack of precise origins, the telescope burst upon the world in the early years of the seventeenth century. On September 25, 1608 (by the Gregorian calendar) the Committee of Councilors of the Province of Zeeland—a lowly tier of local Dutch government—wrote a letter to one of its representatives at the States General— the main governing body—in The Hague. It instructed the delegate to recommend its bearer to Prince Maurice (1567-1625), Count of Nassau, stadtholder of five of the seven Dutch provinces, and commander in chief of the Dutch armed forces. The bearer, they added, claimed to have "a certain device by means of which all things at a very large distance can be seen as if they were nearby, by looking through glasses which he claims to be a new invention." The holder of the letter was a spectacle maker, Hans Lipperhey (1570-1619). Exactly a week later, an entry was made in the minute book of the States General describing Lipperhey's application for a patent.

Little is known about Lipperhey. He was born around 1570 in Wessel, now in Germany, but moved in 1594 to Middelburg, the most important commercial and manufacturing center in southwestern Netherlands. He married the same year and became a Dutch citizen in 1602.

He arrived in The Hague with his telescope—a tube housing two crude, small lenses made of scratched, cloudy glass—a few days before the end of September when the city was hosting a peace

Above *Portrait of Hans Lipperhey who also claimed to have invented the telescope.*

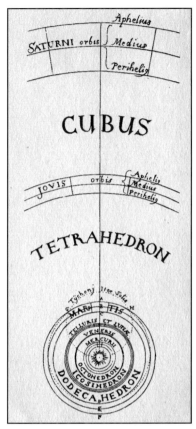

Above *The world system determined from the geometry of the regular solids from Kepler's Harmonices Mundi Libri (Linz, 1619).*

conference. The negotiations between the Dutch and the Spanish over their conflict in the Netherlands had been going on for some time, and because the Dutch had treaties with the English and the French, they too were represented at the conference. Indeed, the French delegation, led by the statesman Pierre Jeannin (1540–1622), acted as mediator between the two hostile parties. Prince Maurice was unhappy with the way the talks were going.

A witness of these events later wrote: "A few days before the departure of [General] Spinola from The Hague, a spectacle-maker from Middelburg, a humble, very religious and God-fearing man, presented to His Excellency [Prince Maurice] certain glasses by means of which one can detect and see distinctly things three or four miles removed from us as if we were seeing them from a hundred paces. From the tower of The Hague, one clearly sees, with the said glasses, the clock of Delft and the windows of the church of Leiden, despite the fact that these cities are distant from The Hague one-and-a-half, and three-and-a-half hours by road, respectively. When the States General heard about them, they asked His Excellency to see them, and he sent them to them, saying that with these glasses they would see the tricks of the enemy. Spinola too saw them with great amazement and said to Prince [Frederick] Henry [Maurice's brother], 'From now on I could no longer be safe, for you will see me from afar.' To which the said prince replied, 'We shall forbid our men to shoot at you.' The master-maker of the said glasses was given three hundred guilders, and was promised more for making others, with the command not to teach the said art to anyone."

Lipperhey was asked to construct binocular telescopes so that one would not have to look through one eye and close the other, to use rock crystal (quartz) instead of glass, and to deliver six instruments within a year. His initial demand for one thousand guilders for each instrument was considered excessive, but accepted. Thus, on October 5, 1608 Lipperhey took his money with the promise of more on the satisfactory completion of the work. The decision on his patent application was postponed. Although the States General had forbidden him to make telescopes for anyone else and obviously desired to keep the invention a secret, Prince Maurice felt no such restriction. Marquis Spinola (1569–1630), the commander in chief of the Spanish forces, who was in

The Hague for the negotiations, had left the city on September 30, after Maurice had shared Lipperhey's secret with him. Undoubtedly, having looked through the device, Spinola had it duplicated by local artisans upon his return to Brussels.

Word traveled fast. Pierre Jeannin (1540-1622), the head of the French delegation and no stranger to intrigue, heard about it and secretly approached Lipperhey asking him to make telescopes for him, but Lipperhey refused. Jeannin however was undeterred. He alerted his spies in Maurice's army and soon found a French soldier who had learned the secret, and sent him back to France with letters for King Henry IV. The telescope was no secret anymore. While Lipperhey's patent request was being considered in Holland, France and Spain had all in various ways learned about the magical lenses.

Below *Compass from the time of Copernicus.*

Curiously, another person was in possession of the same instrument that same autumn. The minutes of the Committee of Councilors of Zeeland for October 14, 1608 state that they interviewed another individual about the telescope. Again, they wrote to their representative in The Hague telling him "...there is here a young man who says he also knows the art, and who has demonstrated the same with a similar instrument." The Councilors' advice was that the secret was out and that the granting of a patent would be superfluous: "We believe that there are others as well, and that the art cannot remain secret at any rate, because after it is known that the art exists, attempts will be made to duplicate it, especially after the shape of the tube has been seen, and from it has been surmised to some extent how to go about finding the art with the use of lenses."

We are not certain of the identity of the second person, but it could have been Zacharias Janssen (1580-1638), then only about twenty years old and, like Lipperhey, a spectacle-maker in Middleburg. In fact, they lived almost next door to each other. Janssen had been born in The Hague, probably the son of a peddler, and when he moved to Middleburg he became a street seller of anything he could get his hands on, only becoming a

Above *Portrait of Zacharias Janssen.*

spectacle-maker somewhat later. He ran up large debts at the local inns, and was often brought before the Court of Justice both because he could not repay his debts and because of his brawling. Though his life was troubled, he had a clever idea. Janssen lived next door to the town's mint and indeed had a cousin who worked there. Soon he entered the counterfeit business. For a time he got away with it thanks to a profit sharing deal with the local sheriff. However, this was not a long-term occupation; the penalty of counterfeiting was death.

Janssen's possible part in the development of the telescope is rather suspicious. His claim could be dismissed as that of an opportunist trying to take the credit for Lipperhey's discovery. This would be a reasonable assumption, were it not for an extra factor that we shall shortly discover.

Perhaps too coincidentally, a man called Jacob Adriaensoon of Alkmaar (a city in the north of Holland), usually known as Jacob Metius (1571–1628), applied for a patent for a telescope the very next day after Janssen's own request. He maintained he had been investigating the power of lenses for two years, and had invented a device for seeing faraway things, which he had now brought to a reasonable state of perfection. His instrument was by all accounts a good one. It seems that Metius had heard about Lipperhey's claim, and hastened to put in his own so as not to be denied what he saw as the rightful fruits of his labor. Consequently, the States General awarded him one-hundred guilders and asked him to improve his telescope. They treated Lipperhey's and Metius' claims equally, except that they had awarded Lipperhey more money. The Councilors were unimpressed with Janssen's claim. On December 15, 1608 the committee further reported that it had examined a binocular telescope made by Lipperhey, but once more rejected his patent application. He was asked to make two more binocular telescopes and given another three-hundred guilders. On February 13, 1609 it was reported that he had delivered them and was given the final three hundred guilders. Neither Lipperhey, Metius, nor

Janssen are mentioned again in the records of the States General.

In Italy, Galileo's friend Paolo Sarpi, the State Theologian of the Venetian Republic and hater of Rome, was still recovering from the brutal attempt on his life by the Pope's assassins. Despite his injuries, which restricted his once regular visits to brothels, he indulged his thirst for learning and took great pleasure in the fact that the Pope hated his correspondence with heretics and intellectuals. Any interesting stranger or traveler to Venice would inevitably meet the diminutive fifty-six-year-old priest with delicate hands, a high brow, and owlish eyes. Some of them would have known that in Rome he was called "the Terrible Friar," and in Rome they would have avoided him, but they had come to upstart, heretical Venice. Sarpi read a pamphlet entitled, "The Embassy of the King of Siam sent to his Excellency Maurice of Nassau," describing the first visit of Siamese to Europe. What he read in the "other news" section of the publication was what really caught his attention. It mentioned the invention of a device by a man from Middleburg that made it possible to see far off things as though near.

At first, Paolo Sarpi was interested but unconvinced: It seemed like old news. Almost everyone who had handled a lens knew of its magnifying effects. He wrote to a friend (below right). Sarpi had also read a recent publication describing a mirror device used by the priest to King Henry IV of France, and at first, he believed that the device from Middleburg also involved a mirror. He may have told Galileo about it, but he was mired in his own problems and it aroused no interest in him.

To complicate matters further, the telescope had actually appeared in public a few weeks before Hans Lipperhey made his patent application in Middleburg, some five hundred kilometers away in Frankfurt, Germany. In his *Mundus Jovialis* (The World of Jupiter) of 1614, German astronomer Simon Marius (1573–1624) revealed hearing about the new instrument for the first time

Above *Portrait of Paolo Sarpi, Galileo's good friend.*

"I don't know if that artisan (from Middleburg) had my idea, or whether the whole thing did not acquire magnification, as rumor always does, over the course of its journey."
Paolo Sarpi

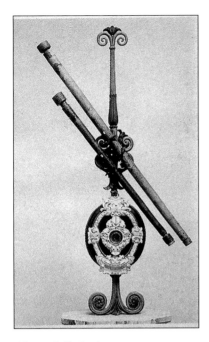

Above *Galileo's telescopes.*

from his patron. Marius had studied in Padua in 1604 and knew Galileo, but Marius was untrustworthy and not averse to plagiarizing Galileo's work. His patron, Johan Philip Fuchs von Bimbach (c. 1525–1626), was at the autumn fair in Frankfurt in 1608. Marius wrote, "...there was then present in Frankfurt at the fair a Dutchman, who had invented an instrument by means of which the most distant objects might be seen as though quite near... Our nobleman [Fuchs von Bimbach] had a long discussion with the Dutch first inventor. At last the Dutchman produced the instrument, which he had brought with him, and one glass of which was cracked, and told him to make a trial of the truth of his statements. So he took the instrument into his hands, and saw the objects on which it was pointed were magnified several times. Satisfied of the reality of the instrument, he asked the man for what sum he would produce one like it. The Dutchman demanded a large price, and when he understood he could not get what he first asked, they parted without coming to terms." That year the autumn fair at Frankfurt began in early September, and lasted about a month. Who exactly was the Dutchman trying to sell a telescope? Some have suggested that it was none other than Zacharias Janssen who, as we have seen, had been a traveling peddler. It is said that upon returning to Middleburg around October 13 from a business trip, Janssen found that Lipperhey had filed a patent, forcing him to do the same.

By the end of 1608, the "art" of the telescope was known in the Netherlands, France, Spain, and Germany, and news of it had arrived in Italy. Small, three-power telescopes began appearing all over Europe. Galileo was bound to see one and recognize its possibilities for science. Or was he?

Galileo was searching for something that would change his fortune. During the latter part of 1608 and for the first half of 1609, he was apprehensive and worried; the anxieties of midlife were threatening to overwhelm him. He was successful, but only moderately so. He had a reputation for an inventive mind, but unconventional views and outlandish tastes. His family duties weighed down on him. His relatives in Florence were nearly destitute and a constant drain on his meager resources. He had to support his mistress and three children, find dowries for his

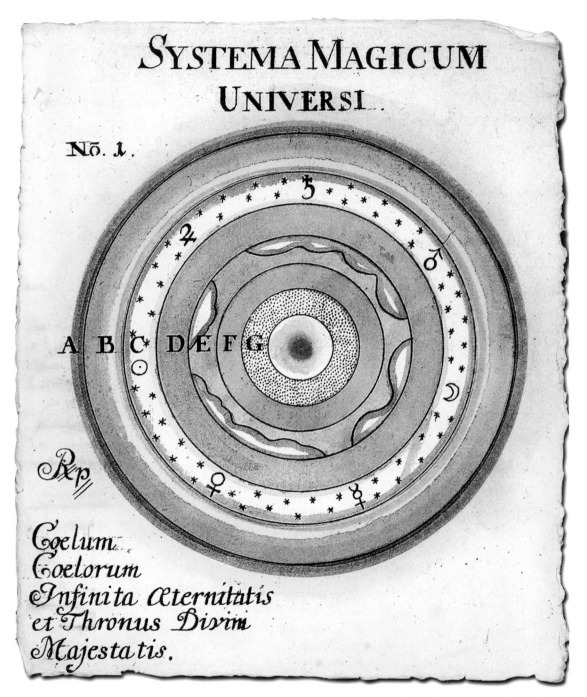

SYSTEMA MAGICUM
UNIVERSI

Nō. 1.

A B C D E F G

℞

Coelum
Coelorum
Infinita Æternitatis
et Thronus Divini
Majestatis.

Above *Ptolemy's geocentric model of the universe, a model studied in the Renaissance.*

Above *Italian astronomers with their instruments, from Ananius' Compotus, printed by Andreas Freitag, Rome, 1493.*

"I have many and diverse inventions, only one of which could be enough to take care of me for the rest of my life...if only I could find a good prince who would like it."
Galileo

sisters, salaries for his servants, as well as money for his ungrateful mother, all on five hundred crowns a year, a respectable, but not handsome salary. This was what not he deserved. His rival, the arrogant and insufferable Cesare Cremonini, was paid twice that. Galileo had complained to the authorities, and had arranged for others to complain on his behalf, but he felt he was in danger of appearing a whining pest. One administrator at the university had already told him that if he did not like what he was being paid, he could always resign and go elsewhere. His dissatisfaction deepened when his university teaching duties increased, as he hated giving lessons. He missed the company of his friend Gianfrancesco Sagredo, and was often bedridden for days at a time with asthma, painful ruptures, and the first worrying signs of arthritis.

He cast a few horoscopes, just to earn additional money, and did some work on the theory of the motion of projectiles through the air, but he knew others were working on the same problems. It was old ground. Galileo was forty-five years old and felt he was going nowhere. He was on the lookout for a big idea, something that would change his destiny and make him famous, and therein lays a mystery.

Was it just bravado that made him claim that he had such a big idea early in 1609? In February, he was talking about a promising new project but was reluctant to give any details. He made it no secret: It was the new thing. He wanted return to Florence and be employed by the Medici Grand Duke: "To serve a Prince. Now that I would love." He shamelessly wrote to the Duke (*left*). Was the telescope Galileo's big idea? It seems certain it was not, at least not initially.

The person who convinced Galileo and Sarpi that the telescope was more important than they thought was Galileo's former pupil Jacques Badovere (1570/1580–c. 1620), who was later thanked in

Galileo's book describing his telescopic observations, the *Siderius Nuncius* (The Starry Messenger). Badovere came from a rich Venetian family that had emigrated to France only to lose their fortune in the St Bartholomew's Day massacre of 1572. Despite this, he was well connected in Paris and kindly disposed to his former tutor. He was convinced that Galileo should take an interest in the telescope, and that he would do so when he fully understood how the instrument had been developed.

Paolo Sarpi suspected that his letters, including those from Paris written by Badovere, were intercepted. His unconventional religious views and opposition to what he saw as the worldly corruption of Rome continued to arouse suspicion, and had already almost cost him his life. The surveillance went to the very top. The nephew of Pope Paul V, Cardinal Scipione Borghese (1576-1633), the secretive and sinister Papal Secretary of State, was reading Sarpi's letters and sending copies to the Pope and the King of France. Thus, the Vatican too heard about the telescope around the turn of the year and became very interested.

Above *Two of Galileo's telescopes now in the Institute and Museum of the History of Science in Florence.*

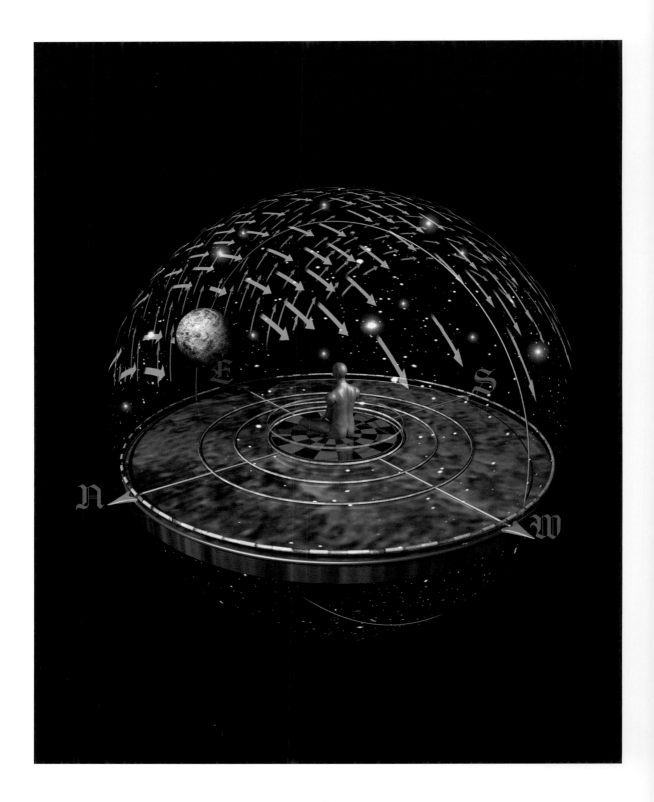

The Hugenot Jerome Groslot de L'Isle (dates uncertain), bailiff of Orléans, had also written to Sarpi about the telescope as early as 1608, and in 1609 wrote again calling its development a "miraculous event." With such praise, Sarpi began to question his own assumption that the instrument was a clumsy and limited device involving a mirror.

Meanwhile Cardinal Borghese moved swiftly to procure a reproduction of the telescope. He contacted another of Galileo's former students, the now loyal Cardinal Guido Bentivoglio (1579–1644), and asked him to acquire the instrument. Thus, the Vatican had one in the summer of 1609, almost certainly a crude device, months before Galileo built one. What the Vatican thought of the possibilities of the new device and its ramifications lies buried in its archives to this day.

That spring news of the invention spread throughout Europe. A crude telescope was soon for sale in the shops of spectacle-makers in Paris by April and in Milan a month later, soon to reach Venice and Naples. Still Galileo did not realize its significance. He had tried to make contact with his former student Badovere late in 1608, but the letter was lost in transit. Sarpi attempted to help the two correspond, but it was only in late March that Badovere finally wrote. Although his first letter contained no details about the Dutch telescope, the second did and it was then that Galileo finally grasped the potential of the telescope and realized he had to move swiftly. There was no time to waste if he was to capitalize on this new invention. It was exactly what he had been waiting for. However, he was not alone in his plans.

In August of 1609, Galileo heard from his friends—the Paduan antiquarian Lorenzo Pignoria (1571–1631) and Paolo Gualdo (1553–1621) from Rome—that a mysterious foreigner had arrived in the city in late July, while Galileo was in Venice visiting friends. The foreigner, probably from the Netherlands, was showing a mediocre telescope. Upon his return to Padua, Galileo was horrified to discover that the stranger had already left for Venice and to hear he was showing the instrument to the authorities and attempting to sell its "secret" to the Venetian government.

Had this excellent opportunity slipped through his fingers? Not long before, toward the end of June, he had moaned to his

Left *Ptolemy's view of the sun, the planets, and the stars have long been discarded, but our perceptions are still Ptolemaic. We look to the east to see the sun rise (when in relation to Earth it is stationary); we still watch the heavens move over us and use the north, south, east, west directions, ignoring the fact that our Earth is a globe.*

Right *Galileo shows his telescope to the Doge of Venice.*

diplomat friend Piero Duodo (dates uncertain) at Padua University about the possibility of improving his position and of increasing his earnings. Duodo wrote to say that he would intercede with the university's authorities, but doubted he would be successful. It seems that Galileo was initially not as interested in the telescope and the possibilities it presented, as to do anything he could to secure favorable attention. Cursing himself for taking so long to be spurred into action, Galileo set to work at once. As he tells us in *The Assayer* published many years later (and in the first draft of the *Starry Messenger*) he succeeded overnight in conceiving and testing a combination of concave and convex lenses to make a telescope. During the next several days, he hardly slept as he busied himself in obtaining or grinding lenses that were more suitable, and constructing a tube for them. On August 20, 1609 he made a payment of twenty-one lire to his instrument maker, and then set out for Venice.

Galileo and Sarpi devised a plan and moved in concert; after all, an improved telescope built elsewhere could arrive in Venice any time. They had to act quickly. It was a device that, according to Galileo was "drawn upon the most recondite speculations in optics." Sarpi's public relations campaign went even further describing it as "one of the fruits of the science which he [Galileo] had professed for more than seventeen years at the University of Padua."

Galileo never met the mysterious stranger who had left the city with his unimpressive telescope that had failed to succeed. However, had it not been for the presence in Italy of an instrument that was causing excitement wherever it was shown, Galileo would most probably have carried his first telescope to Florence, and demonstrated its power to the Grand Duke in order to further his attempts to secure employment there. He could not risk another delay. For all he knew, others would soon duplicate his own achievement. He also feared that by the time he could reach Florence, it would be known there that a similar instrument was being shown in Venice, so that his claim to superiority would evaporate. Thus, from the moment he learned that the stranger had gone from Padua to Venice, he decided to go there himself.

On August 21, Galileo displayed the workings of his telescope from the Tower of Saint Mark to a distinguished group of Venetian gentlemen, including the powerful Doge Antonio Priuli (1548–1623), whom Galileo already knew and who had sent Giordano Bruno to Rome on charges of heresy many years before. Galileo and several worthy patricians climbed to the top of Saint Mark's Basilica. They were greatly impressed by the fact that they could see distinctly the neighboring towns of Treviso and Padua, watch people going in and coming out of a church on the island of Murano, and observe passengers embarking and disembarking from gondolas on the Canal Grande.

Three days later Galileo, capitalizing on the favorable impression his telescope had made, wrote to the Doge saying he wanted to give him the new instrument. He was told to show up at the Senate the next day. When he arrived, the meeting had already started and he took a seat in the waiting room. Shortly thereafter, Priuli, who was in the chair that day, came out to inform him that a motion had been made to double his salary, and confirm his professorship for life. Priuli promised to have the motion voted on immediately.

According to Galileo, the Senate approved it unanimously, but the minutes show that this was not the case: ninety-eight were in favor, eleven against, and thirty abstained. The large number of abstentions may have resulted from the fact that some senators had heard that similar instruments were already on sale. They were generally inferior to Galileo's but they were cheap. Priuli, who called for the vote that raised Galileo's income and gave him life tenure, believed that Galileo had designed a radically new instrument about which he understood the principles. Equally important is the fact that Priuli assumed, along with his fellow senators, that Venice would be given the patent. When this turned out to be false, Priuli felt slighted, and he would be smarting from this for years.

The next thing for Galileo to do was to communicate the news to the Grand Duke Cosimo de' Medici, his natural prince and former pupil, although he realized this could prove embarrassing. Galileo had to explain his failure to return to Florence, and justify the fact that he had given the new and important device to a foreign government as a gift. The recital of events Galileo prepared

Left *Depiction of Galileo's telescope, the book in which he wrote his observations, the Jupiter moons in an orrery, and the planet Jupiter in the distance.*

is plausible, if not precise in all regards. The manner in which he recounted events to the Florentines was conceived to show that he had been the victim of circumstances, and that he had acted under orders of the government that employed him.

"It is six days since I was called by the Signoria," he said, implying that they called him from Padua, whereas on the twenty-third of the month he was already in Venice, and had shown the instrument to others before the rulers called him. On the two successive days it was shown to the select Signoria and then to the whole Senate, he said. The offer made to him was generous, but he had refused it. This was the story written for the eyes of Cosimo. He made a hurried trip to Florence the following month, repaired any damage

Dear and Honored Brother-in-Law:

I did not write after receiving the wine you sent me, for lack of anything to say. Now I write to you because I have something to tell you which makes me question whether the news will give you more pleasure or displeasure, since all my hope of my returning home is taken away, but by a useful and honorable event.

You must know, then, that it is nearly two months since news was spread here that in Flanders there had been presented to Count Maurice a spyglass, made in such a way that very distant things are made by it to look quite close, so that a man two miles away can be distinctly seen. This seemed to me so marvelous an effect that it gave me occasion for thought; and as it appeared to me that it must be founded on the science of perspective, I undertook to think about its fabrication; which I finally found, and so perfectly that one which I made far surpassed the reputation of the Flemish one. And word having reached Venice that I had made one, it is six days since I was called by the Signoria, to which I had to show it together with the entire Senate, to the infinite amazement of all; and there have been numerous gentlemen and senators who, though old, have more than once scaled the stairs of the highest campaniles in Venice to observe at sea sails and vessels so far away that, coming under full sail to port, two hours or more were required before they could be seen without my spyglass. For in fact the effect of this instrument is to represent an object that is, for example, fifty miles away, as large and near as if it were only five.

Now having known how useful this would be for maritime as well as land affairs, and seeing it desired by this Serene Ruler, I resolved on the twenty-fifth of this month to appear in the College and make a free gift of it to His Lordship. And having been ordered in the name of the College to wait in the room of the Pregadi, there appeared presently the Procurator Priuli, who is one of the Governors of the University. Coming out of the College, he took my hand and told me how that body, knowing the manner in which I had served for seventeen years in Padua, and moreover recognizing my courtesy in making such an acceptable gift, had immediately ordered the Honorable Governors [of the University] that, if I were content, they should renew my appointment for life and with a salary of one thousand florins per year; and that since a year remained before the expiration of my term, they desired that the salary should begin to run immediately in the current year, making me a gift of the increase for one year, which is 480 florins at 6 lire 4 soldi per florin. I, knowing that hope have feeble wings and fortune swift ones, said I would be content with whatever pleased His Lordship. Then Signor Priuli, embracing me, said: "Since I am chairman this week, and can command as I please, I wish after dinner to convene the Pregadi, that is the Senate, and your reappointment shall be read to you and voted on." And so it was, winning with all the votes. Thus I find myself here, held for life, and I shall have to be satisfied to enjoy my native land sometimes during the vacation months.

Well, that is all I have for now to tell you. Do not fail to send me news of you and your work, and greet all my friends for me, remembering me to Virginia and the family.

God prosper you.

From Venice, August 29, 1609

Your affectionate brother-in-law

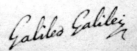

that had been done, and paved the way for negotiations the following spring that eventually culminated in his long-desired appointment by Cosimo. Shortly afterwards he wrote to his brother-in-law (*left*).

Throughout the autumn, Galileo abandoned all his other work, save the lucrative personal tutoring and the occasional horoscope, as he continued to develop his lens grinding and polishing techniques and was soon able to make objective lenses with longer and longer focal lengths. By the beginning of 1610, he was in possession of telescopes that magnified twenty and even thirty times.

It was natural that he would point the telescope to the heavens. The newsletter printed in The Hague in October 1608 that had alerted Sarpi to the existence of the telescope already mentioned that the new device showed stars invisible to the naked eye. However, except for a few additional stars in well-known constellations and star clusters such as the Pleiades, the initial low-powered spyglasses showed nothing new. Such a primitive device could not even show much more detail in the moon than could be seen with the naked eye. Galileo not only managed to make telescopes with much higher magnifications, he also adapted them for astronomical use by supplying his objectives with aperture stops cutting out the light that passed around the edge of the lens, thus increasing the focus. In making these two improvements, he transformed the telescope from a gadget into a scientific instrument.

Galileo's haste had not been misplaced: Others were slowly improving their instruments as well. Thomas Harriot (1560–1621) in England was observing the moon, indeed, mapping it, as early as August 5, 1609 before Galileo had undertaken serious studies of the sky with a telescope that magnified six times.

The heavens were unfolding.

Chapter Four

THE STARRY MESSENGER

With the telescope he developed, Galileo performed observations that allowed him to see the mares, craters, and mounds on the moon, and led him to the discovery of Jupiter's moons, still today known collectively in his honor. In 1610, he was appointed mathematician and philiospher to the Grand Duke of Tuscany. He also studied the peculiar apperance of Saturn and the phases of Venus.

In early October 1608, with his new "perspiculum" in hand, Galileo climbed to the top floor of his house and thrust it through a window that looked out above the rooftops. Smoke rose from many chimneys, and beyond he could see the dome of the Basilica. As the warm and dusky Paduan night fell, a thin, inviting moon hovered in the sky.

He could hardly believe what he saw when he turned his telescope toward the sky: "On the fourth or fifth day after conjunction, when the moon displays herself to us with brilliant horns, the boundary dividing the light from the dark does not form a uniform oval line, but is marked by an uneven, rough, and very sinuous line. For several, as it were, bright excrescences extend beyond the border between light and darkness into the dark part, and on the other hand little dark spots enter into the light." Galileo knew exactly what he was seeing: "...just as the shadows of the earthly valleys diminish as the sun climbs higher, so those lunar spots lose their darkness as the luminous part grows." His telescope showed that the moon was not smooth as Aristotle had stated and as the Church believed.

The modern face of the moon first emerged through Galileo's telescope that autumn evening. It was one of those rare moments when the universe changed: As Galileo watched the moon, the speculations and prejudices of the past were replaced by observation. He would later call the instrument "Old Discoverer," and his telescope can still be seen today at Arcetri near Florence (as

Below *Galileo's drawing of the phases of the moon.*

Above *A photograph of the moon, showing the dark regions Galileo observed through his telescope.*

"Let us speak first about the face of the moon that is turned toward our sight, which, for the sake of easy understanding, I divide into two parts, namely a brighter one and a darker one. The brighter part appears to surround and pervade the entire hemisphere, but the darker part, like some cloud, stains its very face and renders it spotted. Indeed the darkish and rather large spots are obvious to everyone, and every age has seen them. For this reason we shall call them the large or ancient spots, in contrast with the other spots, smaller in size and occurring with such frequency that besprinkle the entire lunar surface, but especially the brighter part. These were, in fact, observed by no one before us."

Galileo

indeed can Galileo's preserved finger that first pointed toward the heavens). He first saw a few irregularities on the moon's crescent face, and made a drawing adding at least five more over the next eighteen days. He looked at the dark spots that had for so long intrigued mankind, the "great or ancient spots," as he called them. Although Galileo could not see deep valleys through his crude telescope, he was convinced they were there, given the pockmarked nature of the surface. He sensed that the moon, too, just like the surface of the Earth itself, could show the variation of lofty mountains and deep valleys. He said the dark spots were "rather even and uniform."

He later wrote: "...if anyone wishes to revive the old opinion of the Pythagoreans, that the moon is another Earth, so to say, the brighter portion may very aptly represent the surface of the land and the darker the expanse of water." He believed that if you could look back at the Earth from the moon's distance, the land would appear bright and the seas dark. Although he was careful not to say there was water on the moon, he certainly hinted at it, writing that the spots are "more depressed." He wrote further: "It is most beautiful and pleasing to look upon the lunar body...from so near... the moon is by no means endowed with a smooth and polished surface, but is rough and uneven and, just as the face of the Earth itself, crowded everywhere with vast prominences, deep chasms, and convolutions."

In his youth, Galileo had contemplated becoming a painter, but Vincenzo forbade it saying there was no money in art, as he well knew. Now, as the first truly modern scientist, Galileo returned to his childhood dream, and painted the moon's portrait in some magnificent drawings. They show the modern moon, a planet stripped of symbolism and myth, a stark world awaiting exploration and discovery. Galileo chronicled his observations of the lunar landscape emerging from darkness.

Several lunar features are quite recognizable in Galileo's lunar drawings. In the second of the series, based on a sketch made on December 3, 1609, the mountains east of Mare Imbrium (Sea of Rains) can be seen as can the sizable crater at the foot of these mountains, probably the Albategnius crater, drawn larger than in reality undoubtedly because of the impression it made on him.

He wrote, "And it is like the face of the Earth itself which is marked here and there with chains of mountains and depths of valleys." Galileo continued: "I render infinite thanks to God for being so kind to make me alone the first observer of marvels kept hidden in obscurity for all bygone ages. I had already ascertained that the moon was a body most similar to the earth, and had shown our Most Serene master as much, but imperfectly, not having such an excellent telescope as I now possess, which, besides showing me the moon, has revealed to me a multitude of fixed stars never yet seen; being more than ten times the number of those that can be seen with the unassisted eye." What he did not know was that Thomas Harriot, the polymath from England, had already observed the moon through a telescope and made his own crude sketches.

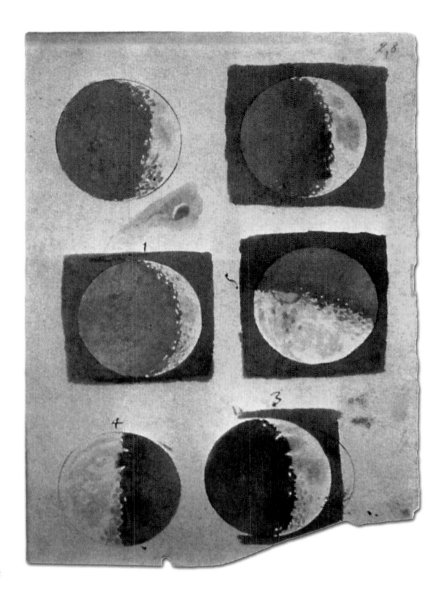

Above *Galileo's drawings of the phases of the moon.*

Galileo visited Florence briefly to show the Grand Duke the moon through his telescope. Cosimo de' Medici, to his great surprise and delight, was able to see that the moon was a body very similar to the Earth. So impressed was he that he gave Galileo eight hundred crowns. Returning home, Galileo continued to improve his telescope by day and turn it toward the heavens by night. He saw the Pleiades, the cluster of stars also known as the "Seven Sisters" the strange misty object in Orion; and observed that just

Above *A photograph showing the Pleiades and stardust.*

about an hour after sunset the planet Jupiter rose in the sky.

| Ori. | * | * | ○ | * | Occ. |

"On the seventh day of the present year 1610, at the first hour of the night, when I inspected the celestial constellations through a spyglass, Jupiter presented himself. And since I had prepared myself a superlative instrument, I saw (which earlier

| Ori. | | ○ | * | * | * | Occ. |

| Ori. | | * | * | ○ | | Occ. |

had not happened because of the weakness of the other instruments) that three little stars were positioned near him—small but very bright. Although I believed them to be among the number of the fixed stars, they nevertheless intrigued me because they appeared to be arranged exactly along a straight line and parallel to the ecliptic, and brighter than others of equal size."

Some regard Galileo's most significant contribution to science his discovery of the four satellites around Jupiter, now collectively named in his honor. He originally thought he saw three stars near Jupiter, strung out in a line through the planet. However, the next evening, these stars seemed to have moved the wrong way. He continued to observe the stars and Jupiter for the next week: "When on January 8, led by some fatality, I turned again to look at the same part of the heavens, I found a very different state of things, for there were three little stars all west of Jupiter, and nearer together than on the previous night. ...My confusion was transformed to amazement," he said. "I have now decided beyond all question that three stars were wandering around Jupiter, as do Venus and Mercury around our Sun."

On January 13, a fourth star appeared. After a few weeks, he had observed that the four never left the vicinity of Jupiter and appeared to be carried along with the planet, and that they changed their position with respect both to each other and to Jupiter. He determined that he was observing not stars, but planetary bodies orbiting around Jupiter. This discovery provided new evidence in support of the Copernican system, and showed that everything did not revolve around the Earth. It was revolutionary. Now he had to put his findings into print before anyone else did.

Above *Observations of the moons of Jupiter, from Galileo's* The Starry Messenger. *The upper sketch shows Jupiter on January 7, 1610, in the midst of three little stars. The planet was then moving west (to the right). In the middle sketch, we see the supposed stars on January 8, with Jupiter to the east (left) of them, quite contrary to what Galileo expected. The lower sketch shows the configuration as it was on January 10. Galileo dedicated nearly half of this little book to the observations of these satellites.*

"...to unfold great and wonderful sights...to the gaze of everyone."
Galileo

He rushed into preparing a book and devoted most of January and February of 1610 to it. He produced watercolors to illustrate his observations, and the *Sidereus Nuncius*, or *Starry Messenger*, appeared the following March (*left*). This is the first true book of modern science, and four hundred years later, it is still fresh, vital, and captivating. Today, it remains fascinating to peer through a telescope with a copy of the Sidereus Nuncius in hand, and to ponder over Galileo's words while looking at the moon. All astronomers and students of this science should read it, as they would find the passion and the sense of discovery with which it was written inspiring.

He wrote in the book: "I should disclose and publish to the world the occasion of discovering and observing four planets, never seen from the beginning of the world up to our own times, their positions, and the observations made during the last two months about their movements and their changes of magnitude; and I summon all astronomers to apply themselves to examine and determine their periodic times, which it has not been permitted me to achieve up to this day. ... I therefore concluded, and decided unhesitatingly, that there are three stars in the heavens moving about Jupiter, as Venus and Mercury around the Sun; which was at length established as clear as daylight by numerous other subsequent observations. These observations also established that there are not only three, but four, erratic sidereal bodies performing their revolutions around Jupiter."

Within hours of the publication of the *Sidereus Nuncius* on March 12, 1610, the British Ambassador Sir Henry Wotton had obtained a copy and sent it post haste to the court of King James. Galileo's sensational work quickly reached Germany, where it was reissued in a pirated edition in Frankfurt.

In the haste, little time or expense was devoted to copying Galileo's careful lunar engravings. Consequently, the Frankfurt edition contains woodcuts that were much less skillfully prepared than the original engravings. In addition, they were oriented incorrectly and identified wrongly. The Frankfurt woodcuts were the source for the illustrations in most of the later editions of the *Sidereus Nuncius*, leading unwary scholars to criticize Galileo's superb renditions unfairly as crude and unrealistic.

Above *Family portrait of Jupiter's Great Red Spot and the Galilean Satellites.*

Ever the publicist, Galileo was less than modest about his achievements (*right below*).

What better way to impress the Grand Duke, Galileo thought, than to dedicate the four stars circling Jupiter to his eminence? "Behold four stars have been reserved to bear your famous name. These bodies belong not to the inconspicuous multitude of fixed stars, but to the brighter ranks of the planets. They move above noble Jupiter as children of his own. Jupiter! Benign star! Next to God, the source of all good things! At the instant of your Highness' birth, he merged from the turbid mists of the horizon, illuminating the eastern sky from his own royal house. He looked out from that exalted throne upon your auspicious birth, pouring forth all his splendor and majesty, so that your tended body and mind might imbibe that universal influence and power."

The *Starry Messenger* took Europe by storm. All 550 copies printed were sold within a week. Here was the new Christopher Columbus of the heavens. Requests for telescopes arrived from all over Europe and Galileo was not shy in sending them out with letters praising himself, as this example, a letter to the Prince of Venice, shows:

"Most Serene Prince, Galileo Galilei most humbly prostrates himself before Your Highness, watching carefully, and with all spirit of willingness, not only to satisfy what concerns the reading of mathematics in the study of Padua, but to write of having decided to present to Your Highness a telescope ('Occhiale') that will be a great help in maritime and land enterprises. I assure you I shall keep this new invention a great secret and show it only to Your Highness. The telescope was made for the most accurate study of distances. This telescope has the advantage of discovering the ships of the enemy two hours before they can be seen with the natural vision and to distinguish the number and quality of the ships and to judge their strength and be ready to chase them, to fight them, or to flee from them; or, in the open country, to see all details and to distinguish every movement and preparation."

However, from the beginning, there was opposition to these new ideas, isolated at first but it grew. There were those who did not look through the new telescope correctly and there were those who did not want to see. A letter from German Lutheran Martin Horky (dates uncertain) to Johannes Kepler in April, 1610 said:

"To apply oneself to great inventions, starting from the smallest beginnings is no task for ordinary minds. To divine that wonderful arts lie hidden beneath trivial and childish things is a conception of superhuman talents."
Galileo

"Galileo Galilei, the mathematician of Padua, came to us in Bologna and he brought with him that spyglass through which he sees four fictitious planets. On the twenty-fourth and twenty-fifth of April, I never slept, day and night, but tested that instrument of Galileo's in innumerable ways, in these lower as well as the higher [realms]. On Earth it works miracles; in the heavens it deceives, for other fixed stars appear double. Thus, the following evening, I observed with Galileo's spyglass the little star that is seen above the middle one of the three in the tail of the Great Bear, and I saw four very small stars nearby, just as Galileo observed about Jupiter. I have as witness most excellent men and most noble doctors, Antonio Roffeni (1580–1643), the most learned mathematician of the University of Bologna, and many others, who with me in a house observed the heavens on the same night of the twenty-fifth of April, with Galileo himself present. But all acknowledged that the instrument deceived. And Galileo became silent, and on the twenty-sixth, a Monday, dejected, he took leave from Signor Magini (1555–1617) very early in the morning. And he gave no thanks for the favors and the many thoughts, because, full of himself, he hawked a fable. Signor Magini provided Galileo with distinguished company, both splendid and delightful. Thus the wretched Galileo left Bologna with his spyglass on the twenty-sixth. Unknown to anyone, I have made an impression of the spyglass in wax, and when God aids me in returning home, I want to make a much better spyglass than Galileo's."

Francesco Sizzi (dates uncertain), an astronomer from Florence, also argued against Galileo's discovery of four moons orbiting around Jupiter, with comments typical of the time saying that just as there are seven "windows" in the head (two nostrils, two eyes, two ears, and a mouth), so in the heavens God had placed two beneficent stars (Jupiter, Venus), two maleficent stars (Mars, Saturn), two luminaries (sun and moon), and one indifferent star (Mercury). The seven days of the week followed from these, he added. "If we increase the number of planets, this whole system falls to the ground. Moreover, the satellites are invisible to the naked eye and therefore can have no influence on the earth, and therefore would be useless, and therefore do not exist," concluded Sizzi.

Above *Galileo showing people the moons of Jupiter orbiting around the massive planet.*

The German astronomer Simon Marius (1573–1624) claimed to have observed Jupiter's moons as early as late November 1609 (about five weeks prior to Galileo) and had begun recording his observations in January 1610 at about the same time Galileo was first making his. However, since Marius did not publish his observations right away as Galileo had done, his claims were impossible to verify, and besides he had been shown to be untrustworthy on other scientific matters. Since Galileo's work was more reliable and extensive, he was given the credit for discovering the moons of Jupiter. However, in 1614, Marius provided the individual names of the moons as we know them today, based on a suggestion from Johannes Kepler: "Jupiter is much blamed by the poets on account of his irregular loves. Three maidens are especially mentioned as having been clandestinely courted by

Jupiter with success. Io, daughter of the river Inachus; Callisto of Lycaon; Europa of Agenor. Then there was Ganymede, the handsome son of King Tros, whom Jupiter, having taken the form of an eagle, transported to heaven on his back, as poets fabulously tell. ...I think, therefore, that I shall not have done amiss if the First is called by me Io, the Second Europa, the Third, on account of its majesty of light, Ganymede, the Fourth Callisto. This fancy, and the particular names given, were suggested to me by Kepler, Imperial Astronomer, when we met at Ratisbon fair in October 1613. So if, as a jest, and in memory of our friendship then begun, I hail him as joint father of these four stars, again I shall not be doing wrong."

Galileo planned a change of employment: "It is impossible to obtain from a Republic, however splendid and generous, a stipend without duties attached to it; for to have anything from the public one must work for the public, and as long as I am capable of lecturing and writing, the Republic cannot hold me exempt from duty, while I enjoy the emolument. In short, I have no hope of enjoying such ease and leisure as are necessary to me, except in the service of an absolute prince."

"I desire greatly," he wrote to the Grand Duke's secretary, "to have my mind set at rest on that business which we discussed lately at Pisa. For, seeing that every day is one more day gone, I am entirely resolved to fix once for all on the mode in which the rest of my life is to be passed, and turn all my energies to bring to a termination the labors of all my past life, from which I hope to gain some renown." After mentioning his income and prospects at Padua, he continues: "And, in short, I should wish to gain my bread by my writings, which I would always dedicate to my Serene Master. Of useful and curious secrets I possess so many, that their very abundance does me harm; for if I had but one, I should have esteemed it greatly, and perhaps, through it, I might have found that fortune which as yet I have not met with, nor have I sought it but they are no good to me, or rather they can be no good except to princes; for they alone make war, erect fortresses, and for their royal pleasure spend such sums of money as private gentlemen cannot, any more than I can. The works which I wish to finish are principally these: two books on the system of the universe; an

Left *Our solar system.*

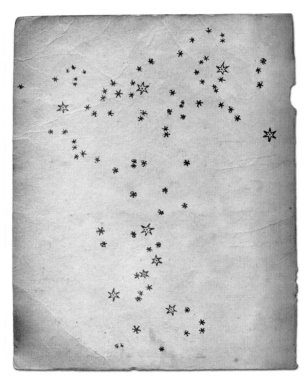

Above *Galileo's sketch in* The Starry Messenger *of the stars in Orion's belt (the three large stars across the top of the figure) and sword. Most of the smaller stars in this drawing were being seen for the first time since creation. Curiously, he does not depict the great nebula. This may be because he had seen it but could not 'resolve' it into stars (as he did the Milky Way), and so decided to postpone mention of it until some future date, when he hoped to have telescopes powerful enough to show the component stars.*

"...the smallest of them... when observed through the telescope, can scarcely be perceived, and only with fatigue and injury to the eyes."
Galileo

immense work full of philosophy, astronomy, and geometry: three books on local motion—a science entirely new—no one, either ancient or modern, having discovered any of the marvelous accidents which I demonstrate in natural and violent motions; so that I may with very great reason call it a new science."

In July 1610, all his hard work and flattery paid off when he was appointed Chief Mathematician and Philosopher to the court of the Grand Duke of Tuscany with an astonishing salary of 1,000 crowns a year. However, not everyone was jubilant. Those he was leaving behind in Venice were bitter. They felt they had claim on his services, because they were first to patronize him and his telescope. That he should sell out to the highest bidder disgusted them. Antinio Priuli, the Doge of Venice who had helped Galileo, said he hoped that he would never again set eyes upon him.

From Florence, Galileo saw other wonders in the sky. His early observations of the sun were made while looking at the star through thick clouds at sunset. For hundreds of years spots on the sun had been seen from time to time, but no one was sure what they were or even if they were real. Galileo saw them (*below left*).

He was lucky not to injure his eyes permanently. In the Florence History of Science Museum, his instruments have been measured and tested and the brightness of the images seen through them can be estimated. One of Galileo's fourteen-power telescope provided an image of the sun forty percent less bright than on the retina of the unaided eye. A twenty-one-power telescope showed an image at sixty percent of the brightness on the naked eye. The sun's brightness had been reduced by the poor quality, uncoated glass, and misalignment of the lenses in his telescopes. Since his observations were made at sunset through clouds, it appears he did not injure his eyes from focused sunlight. He was fortunate.

Galileo soon adopted the method of telescopic projection onto a screen as soon as he learned about it from his former pupil Benedetto Castelli in May 1612. Castelli (1578-1643) was a

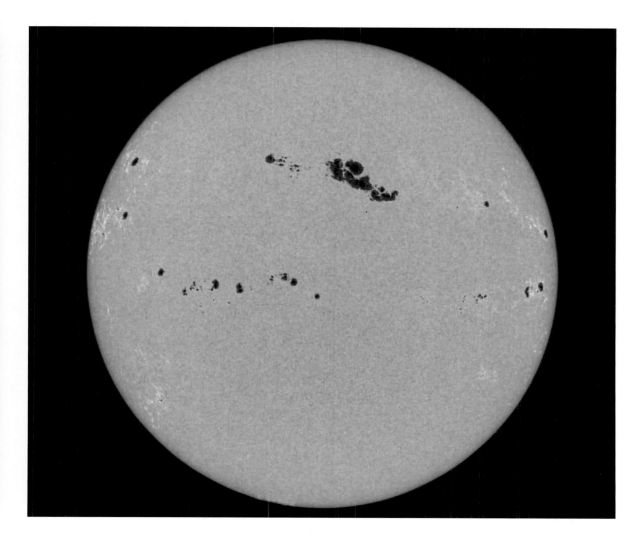

Above *Photograph showing sunspots as magnetic depressions.*

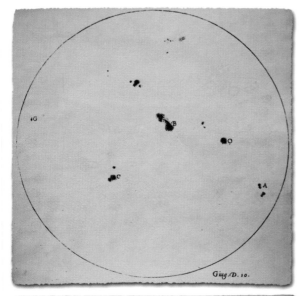

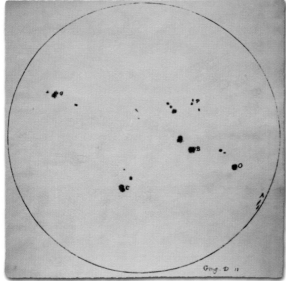

Above *Drawing of sunspots on June 11 and 12, 1612, from Galileo's* Letters on Sunspots. *Under suitable conditions, sunspots can be seen with the naked eye. The first telescopic observation of a sunspot was made in England by Thomas Harriot (c. 1560–1621) in late 1610.*

student of Galileo, and wrote to him about the projection techniques for observing the sun, including solar drawings made by projection onto a circle of standard diameter. This was not only less damaging than viewing with the naked eye, but it also improved scientific study as drawings could be stored and compared. Castelli, beginning collaboration with Galileo, started keeping an accurate record of the movement of sunspots, and, dividing the solar disk into fifteen segments, he showed the trajectory of sunspots tracking them with progressive measurements. These accurate measurements allowed Galileo to demonstrate that sunspots were at or near the solar surface, thus ending years of speculation and controversy.

In July he observed Saturn. Perhaps he was hoping that it would be like Jupiter, but it was not, and Galileo was mystified. Of course, he had no idea that the planet was surrounded by a ring system. He wrote to the Duke saying, "The planet Saturn is not alone, but is composed of three, which almost touch one another and never move nor change with respect to one another. They are arranged in a line parallel to the zodiac, and the middle one (Saturn itself) is about three times the size of the lateral ones." He also described Saturn as having "ears." In 1612, the plane of the rings was oriented directly at the Earth so they appeared to vanish. Puzzled, Galileo wondered, "Has Saturn swallowed his children?" referring to the myth of the God Saturn eating his own children to prevent them from overthrowing him. Then, in 1613, they reappeared again, further confusing Galileo.

It was in 1655 that the Dutch astronomer Christiaan Huygens (1629-1695) suggested for the first time that Saturn was surrounded by a ring. Using a telescope far superior to those available to Galileo, Huygens observed Saturn and wrote that, "It is surrounded by a

Right *Harmony of the spheres. The Church believed the heavens to be ruled by the divine order, not by scientific principles.*

thin, flat, ring, nowhere touching, inclined to the ecliptic."

By now, Galileo had settled in Florence and put the annoyance felt in Venice at his departure behind him. He continued to nurture the interest of the wealthy and influential, among whom a well-heeled young nobleman called Filippo Salviati (1582–1614). His villa, fifteen miles out of Florence, became Galileo's second home.

Castelli wrote to Galileo in December 1610 pointing out that if the Copernican system were true, then it would be possible to observe phases on Venus like those seen on the moon. Although in 1610 Venus had not been entirely visible until October, Castelli had observed some movements. On December 11, Galileo sent an

Below *A navigator learns how to use a cross-staff to measure the angle between the moon and a particular star. Elsewhere in this woodcut of 1533, surveyors are using the instrument to measure the height of a tower. To use a cross-staff to measure the angle separating two objects, the observer slides the cross bar towards or away from himself until each object is in line with one or the other end of the bar; the angle can then be read off from the scale on the central member.*

anagram to Kepler in Prague on Castelli's annotations. Anagrams were a technique used at the time to record a discovery without revealing it to the public. Kepler was to be its keeper as he was not yet given the key for deciphering the message. The observations of Venus finally convinced Galileo that Copernicus was correct. With this new information, in January 1611 he wrote to the Secretary of State and to the Grand Duke Cosimo in Pisa suggesting a rendezvous in Rome so that he could reveal his new discovery. In March, he set off observing the satellites of Jupiter each evening during the six-day journey from Florence.

Christopher Clavius, by now an old man with only a year to live, remembered receiving the young Galileo twenty-four years earlier. Galileo wrote him of his observations. Although Clavius was impressed, Galileo knew he was too old to change, and that he would cling to Aristotle, saying that to see the satellites of Jupiter men had to make an instrument that would create them. Clavius bravely acknowledged that things were changing, and that future generations would have to find new formulas. "The whole system of the heavens is broken down and must be mended," he said.

Cardinal Camillo Borghese became Pope Paul V in 1605, elected over a number of valid contenders, including Roberto Bellarmine. The fact that he had appeared to maintain his neutrality in the factional politics of Rome, had made Borghese an ideal compromise candidate. Rather chubby in appearance, he was very stern and stubborn in character, behaving as the lawyer he was. His first act was to send the bishops who were living in Rome home to their bishoprics, for the Council of Trent (1545–1563) had insisted that every bishop live in his diocese. Once in power, the Pope who had been elected as a compromise began to exhibit eccentric tendencies.

During Pope Paul's papacy the conflict between Venice and Rome intensified, and he was thought to have plotted the assassination of troublesome Paolo Sarpi. The Pope's chief advisor was Cardinal Bellarmine, the Inquisitor of Giordano Bruno, who had treaded a dangerous course when he helped save Sarpi's life by warning him of an impending attack.

In his meeting with Pope Paul V, Galileo was allowed to stand up. After he had listened attentively, the Pope said he had his support. Later, he changed his mind.

As Galileo grew in popularity in Rome, Bellarmine decided to investigate the implications of the telescopic claims. He said he had seen some wonderful things through one, but had also heard conflicting opinions. Bellarmine wrote to Clavius and other mathematicians at the Collegio Romano for advice. In a letter sent by Clavius to Bellarmine, he stated: "...it appears more probable that the [moon's] surface is not uneven, but rather that the lunar body is not of uniform density and has denser and rarer parts, as are the ordinary spots seen with the natural sight." The idea that the moon really was perfect beneath some vapors was old and Galileo's observations demolished it. Other things made Bellarmine uneasy. When Galileo visited the Collegio Romano he was treated with polite caution by the academic staff, although the students cheered him.

While in Rome, Galileo was asked to join a private science club that had been founded in 1603 by Federico Cesi (1585–1630), an aristocrat from Umbria, the son of the Duke of Acquasparta, and a member of an important family. Cesi was passionately interested in natural history, above all botany. This was the first academy of sciences to exist in Italy, and it subsequently acted as a locus for the scientific changes of the time. It was named after the lynx, an animal whose sharp vision symbolized the prowess of observation required by science. In 1871, it became Italy's official scientific academy.

Cesi started the Accademia dei Lincei with three of his friends: the Dutch physician Johannes Van Heeck (1579–?), and two fellow Umbrians: mathematician Francesco Stelluti (1577–1653) and polymath Anastasio de Filiis (1577–1608). Cesi wanted them to investigate all natural sciences. He planned to experiment freely and to observe phenomena honestly, respectful of tradition, but untrammeled by blind obedience to authority, even that of Aristotle and Ptolemy, whose theories the new science called into question. The academy's motto, chosen by Cesi, was "Take care of small things if you want to obtain the greatest results."

Galileo was inducted to the exclusive academy on December 25, 1611 and became its intellectual leader. Becoming a member of the academy was such an honor, that he signed himself Galileo Galilei Linceo thereafter. The academy published his works and later supported him through his disputes with the Roman Catholic Church. Cesi hosted a lavish banquet for Galileo two weeks after he arrived in Rome. Among the guests was a Greek mathematician called John Demisiani (dates uncertain), who that evening suggested that Galileo should call his instrument a "telescope."

Galileo won the confidence of Rome and then, in a masterpiece of showmanship, unveiled his latest discovery—sunspots—something he had not yet revealed publicly. Although he had shown sunspots to a number of people during his visit to Rome, with all the diversions in the city, he did not undertake a careful study until the following year.

Sunspots are dark blemishes on the sun's surface, sometimes big enough to be seen with the naked eye. Records of sunspot observations in China go back to at least 28 BCE. In the West, it is possible that the Greek philosopher Anaxagoras observed one in 467 BCE. In Aristotelian cosmology, the heavens were perfect and unchanging, and therefore sunspots were not taken into consideration. A very large spot seen for no less than eight days in 807 BCE was simply interpreted as a passage of Mercury in front

Below *A photograph showing the pock-marked surface of the moon.*

of the sun. In 1607, Johannes Kepler wanted to observe a predicted transit of Mercury across the sun's disk, and on the appointed day he projected the sun's image through a small hole in the roof of his house (thus forming a camera "obscura") and when he saw a black spot he interpreted it as Mercury. Had he persisted on his observation the next day, he would have seen the spot again. Knowing that Mercury only took a few hours to cross the sun's disk during one of its infrequent transits, Kepler would have realized that what he observed could not have been Mercury.

Galileo wrote of sunspots: "I suspect that this new discovery will be the signal for the funeral, or rather for the last judgment, for pseudo-philosophy. The dirge has already been heard in the moon, the Medicean stars, Saturn, and Venus. And I expect now to see the Peripatetics [members of a school of philosophy in ancient Greece that had a revival in late medieval Europe and promoted Aristotle's doctrines] put forth some grand effort to maintain the immutability of the heavens."

In the meantime, Christoph Scheiner (1575–1650), a Jesuit priest and scientist from Germany, began his study of spots in October

Below *Christoph Scheider measuring sunspots with a Keplerian telescope. In 1611, Scheider's skeptical religious superiors had considered his discovery of sunspots so implausible that its announcement would bring ridicule on the Society of Jesus, and at first he was required to publish anonymously. Unlike Galileo, Scheider persevered for many years in his sunspot studies, which culminated in his massive Rosa ursina, published between 1626 and 1630. On the wall in the picture hangs an astrolabe, an instrument of a bygone age.*

1611 using a telescope with filters of colored glass. His first tract on the subject, *Tres Epistolae de Maculis Solaribus Scriptae ad Marcum Welserum* (Three Letters on Solar Spots written to Marc Wesler) appeared the following January. Welser was a scholar and banker in Augsburg, and in turn a patron of local scholars. Schneider's book was the beginning of a debate with Galileo over sunspots. Scheiner believed, as did both Aristotle and the Catholic Church, that the sun was perfect and unblemished. He argued therefore that sunspots must be satellites of the sun. Welser invited Galileo to comment on Scheiner's writings, and Galileo responded with two letters to Welser stating that sunspots are on or near the surface of the sun, that they change their shapes, and that they are often seen to originate on the solar disk and perish there, leading to the conclusion that the sun is not perfect.

Meanwhile, the brooding resentment among Galileo's enemies resulted in their attempt to topple him from his pedestal. Led by Lodovico delle Colombe (1565–1616), one of Galileo's adversaries and a staunch Aristotelian philosopher, his rivals sought to ambush Galileo in a conflict between science and Scripture knowing it was a battle Galileo could not win. Colombe, aware that he was no match for Galileo on an intellectual level and who had already been embarrassed in a dual of wits with him, was intent on defaming Galileo any way he could, and was the first to use the Bible as a weapon against the scientist. He circulated his treatise called "Against the Motion of the Earth" in manuscript form, and the work quoted several examples from Scripture that contradicted the Copernican system. One such reference, Joshua 10:12-13, used time and again, became closely identified with the disagreement: "Thus, Joshua, after defeating the Philistines, commanded the sun to stand still, which implied that the sun usually moves."

Colombe persuaded a Dominican priest, Tommaso Caccini (1574–1648), to deliver a sermon strongly denouncing Galileo and his followers, and condemning the notion of a moving Earth. Caccini had entered the Dominican order as a teenager, beginning his career in the monastery of San Marco in Florence, and became gradually renowned for his passionate sermons. Caccini had never liked Galileo, and now he had a reason to tell his congregation why. There was more than just dislike in his words. He was a frustrated

Dominican friar, and hoped the fuss he was about to cause would ensure that the ungrateful Church hierarchy would notice his talents at last. It has been said of him that "his fanaticism was never divorced from personal ambition for advancement within the Dominican order."

On December 21, 1614 the forty-year-old Friar Caccini gave a sermon at the church of Santa Maria Novella in Florence targeting Galileo. He lit a fuse. His thesis was based on the book of Joshua from which he quoted the following passage:

And then Joshua said in the sight of Israel:
Sun, stand thou still upon Gibeon;
And thou Moon, in the valley of Ajalon.
And the Sun stood still, and the moon stayed,
Until the people had avenged themselves
Upon their enemies
So the sun stood still in the midst of heaven
And hastened not to go down a whole day
And there was no day like that before or after it
That the lord harkened unto the voice of man.

Although the text of the rest of the sermon has been lost, historians suggest that Caccini preached that mathematics and science were contrary to the word of the Bible, and therefore sacrilegious, singling out Galileo and his followers as examples of heretics. He quoted from the first chapter of the Acts of the Apostles: "O ye men of Galilee, why stand ye gazing up into heaven." He contrasted Galileo's allegedly heretical acts to the unwavering faith of the inhabitants of Galilee, and called for the banishment of mathematicians from Christian States.

Caccini's behavior, as some scholars have pointed out, "stands out in contrast to that of nearly all the other churchmen involved" in the subsequent controversy. Father Luigi Maraffi (dates uncertain), the Dominican Preacher General and a supporter of the Copernican theory, wrote a letter of apology to Galileo: "Unfortunately, I have to answer for all the idiocies that thirty or forty thousand brothers may and do actually commit." Maraffi had also made it clear to his Dominican friars that he would not tolerate this kind of behavior: "We should not open the door for every impertinent individual to come out with what is dictated to

Above *The great point of contact between astronomy, astrology, and religion was always the fact that heaven is both physical and a spiritual realm; the properties and the mysteries of the heavens may be studied by the scientist and the mystic. This raised obvious questions about the nature of the stars: what actually were they? As physical bodies in the heavens, how did they relate to the divinities who were also supposed to dwell there? The nature of the stars had been a matter of speculation for Mesopotamians, Persians, and Greeks, and the conclusion that the stars were divinities, animate and intelligent, fuelled the growth of astrology.*

Dies diei eructat uerbum
Numerus
Mensura
Pondus
Videbo Cælos tuos, opera digitor. tuor.
Non Inclinabitur in seculum seculi
Ponderibus librata suis
F. Curtius Bon. Incid.

him by the rage of others and by his own madness and ignorance." The severest rebuke came from Caccini's own brother, "It was a silly thing to get him embroiled in this business. What idiocy is this of being set bellowing at the prompting of those nasty pigeons. This performance of yours makes no sense in heaven and earth. I who am no theologian can tell you what I am telling you, that you have behaved like a dreadful fool."

Galileo responded by writing a letter arguing that nothing he believed was in conflict with Scripture. In his opinion, Scripture dealt with natural matters cursorily and elusively, reminding us that its proper concern was not natural phenomena, but the soul of man. According to Galileo, Scripture adjusted its notions regarding nature to the simple minds of ordinary people. He argued that Christian sacred writings were not intended to validate science, and defended this point by quoting Cardinal Baronius (1538–1607), a well-known ecclesiastical historian, who had remarked that, "The Holy Ghost intended to teach us how to go to heaven, not how the heavens go."

Although Caccini's attacks were excessive and irrational, they struck a chord. The stage had been set for a dramatic confrontation with Rome. Although Galileo and many members of the Church did their best to avoid the collision, it seems, however, no one had the power to prevent it.

Left The frontispiece to the massive compendium of astronomical knowledge, Almagestum Novum (1651), by the Jesuit astronomer Giambattista Riccioli (1598–1671). Riccioli was one of a number of Tycho's contemporaries and successors to propose planetary systems in which the Earth (in some cases spinning daily) was at the center orbiting the sun, carrying all or some of the lesser planets with it. In Riccioli's system, Venus, Mercury, and Mars were satellites of the sun, while the moon, the sun, Jupiter, and Saturn orbited the Earth. Urania weighs this system against that of Copernicus, and finds Riccioli's the winner. Ptolemy's system lies discarded while its author, reclining on the ground, expresses his admiration for the corrections made to it. On the left, is Argus Panoptes ("all-seeing"), who in Greek mythology had eyes all over his body and so symbolized the starry heavens. The various phrases are from the Vulgate translation of the Psalms, while at the top, as at Belshazar's feast in the Book of Daniel, a disembodied hand writes "Number. Measure. Weight."

Chapter Five

THE FACE OF THE INQUISITION

In 1611, Galileo went to Rome, where he was invited to join the prestigious Accademia dei Lincei. While in Rome he also observed sunspots. In 1612, opposition arose to the Copernican theories that Galileo supported. In 1614, from the pulpit of Santa Maria Novella in Florence, Father Tommaso Caccini denounced Galileo's opinions on the motion of the Earth, judging them dangerous and close to heresy. Galileo returned to Rome to defend himself against these accusations. However, in 1616 Cardinal Bellarmine personally handed him an admonition enjoining him to neither advocate nor teach Copernical astronomy because it was contrary to the accepted understanding of the Holy Scriptures.

After his ordeals in Rome, Galileo returned to Florence tired and unwell with rheumatic pains. He took to his bed for days to recover, and remained unaware of the resentment slowly brewing against him. He sought the company of his friend Filippo Salviati. With him and two mischief-making professors from the University of Pisa, Galileo was soon embroiled in a debate about ice. One of the professors said that ice was condensed water; therefore, because it was heavier than water it should sink. Galileo said that he thought it was the other way round: Ice, he maintained, was rarified water, therefore lighter and should always float. The professors said that the only reason ice was seen to float on frozen ponds was that it was thin and broad. Galileo disagreed.

A few days later, the troublesome Lodovico delle Colombe, Galileo's enemy who had convinced Caccini to deliver his condemning sermon against Galileo, heard of the debate and, seeing yet another opportunity to humiliate Galileo, offered a contest. Colombe would show by experiment that the sinking of objects in water depended on their shape. He invited Galileo to prove otherwise. Canon Francesco Nori (dates uncertain), a friend to both contestants, was to act as referee.

"Signor Ludovico delle Colombe," Galileo wrote, "believes that shape affects solid bodies with regard to their sinking or floating in a given medium such as water. I, Galileo Galilei, deem this not to be true. I affirm that a solid body of spherical shape which sinks to

Below *Art, mathematics, and religion were deeply connected, as depicted in this painting by Jacopo de' Barbarai (1440–1515).*

Left *Tuning fork in water showing the force of sound vibration.*

the bottom will also sink no matter what its shape is. I am content that we proceed to make experiments of it."

During the following days, Galileo was hard at work in his workshop preparing a variety of materials in various shapes and sizes. He determined that the best way to see if an object would float on water was to hold it under the surface and then release it to see if it emerged again. All was set for the confrontation, but at the appointed date delle Colombe failed to turn up. Undeterred, Galileo wrote him suggesting a new date to meet at his friend Francesco Salviati's house in Florence. Everywhere in the city people were gossiping about the impending contest. It came to the ears of Cosimo de'Medici who was not keen for Galileo, whom he patronized and protected, to suffer a public defeat. He suggested that instead of competing Galileo write his theory in book form in a fashion that befitted his dignified position associated with the Grand Duke. Thus, at the next appointed time, Galileo refused to take part in the experiment saying that he would write all his arguments instead. It appeared to some that Galileo had lost face, but the treatise he produced and dedicated to Grand Duke Cosimo was a masterpiece. *Bodies in Water* was Galileo's first published work on physics—perhaps not as dramatic as *The Starry Messenger* but it represented thoroughly researched science.

At the end of September 1614 Cardinal Maffeo Barberini (1568-1644), son of a wealthy Florentine family and later to become Pope Urban VIII, whom Galileo had met during his visit to Rome earlier in the year, paid a visit to Florence and saw the various demonstrations of floating bodies Galileo had devised. He was impressed, but despite their apparent friendship it would be Barberini who would summon Galileo to Rome many years later to face the Inquisition. On the day he saw the experiment, Barberini, like everyone else, was convinced that Galileo was right.

Many are the reasons, most Serene Lord, for which I have set myself to the writing out at length of the controversy which in past days had led to so much debate by others. The first and most cogent of these was your hint, and your praise of the pen as the unique remedy for purging and separating clear and sequential reasoning from confused and intermittent altercations in they who especially defend the side of error, on one occasion noisily deny that which they had previously affirmed, and on the next, pressed by the force of reason, attempt with inappropriate distinctions and classifications, cavils, and strained interpretations of words to slip through one's fingers and escape by their subtleties and twistings about, not hesitating to produce a thousand chimeras and fantastic caprices little understood by themselves and not at all by their listeners.

Galileo Galilei

Above *A woodcut from the book*
L'atmospehere mètéorologique populaire by
Camille Flammarion (1841–1925) shows
someone breaking through the medieval
world to see the underlying mechanism
that turns the world.

Above *Portrait of Maffeo Barberini, who became Pope Urban VIII.*

Taking advantage of the presence of the increasingly important Cardinal Barberini, Cosimo staged a replay of the floating bodies, without delle Colombe. Also in attendance was the Duke's mother, the Grand Duchess Christine. In front of the audience, Galileo held a large block of ice under water and released it theatrically. When it emerged he accepted the applause, and felt his reputation had been restored after failing to contest directly with delle Colombe earlier. However, the strain of public life was beginning to take its toll, and he was ill for two months following the experiment. His friends said he never had a truly healthy day in any city. He suffered from arthritis, pains in his kidneys and chest. He told the Duke, "These things disturbed my mind and made me melancholy, which in turn augmented them; yet I have done something, haltingly, in a few days I will send you a Discourse on a certain dispute with some Peripatetics. That sent, I want to spend a few days replying to letters, meanwhile not neglecting celestial observations."

Years earlier, Galileo had studied the activity of sunspots. The popular opinion continued to be that the spots were moving stars, but Galileo was convinced they were real blemishes on the solar surface. Now, at dawn and dusk from Salviati's villa, he observed them as they appeared and disappeared and moved across the face of the sun. In May he wrote to Federico Cesi, the founder of the Accademia dei Lincei;

"I have finally concluded, and I believe I can demonstrate necessarily, that they are contiguous to the surface of the solar body, where they are continually generated and dissolved, just like clouds around the earth, and are carried around by the sun itself, which turns on itself in a lunar month with a revolution similar (in direction) to those others of the planets, that is, from west to east around the poles of the ecliptic; which news I think will

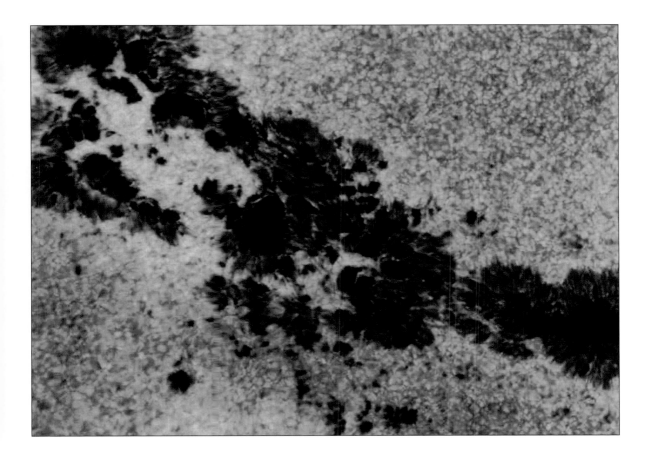

Above *Sunspots show as magnetic depressions.*

be the funeral, or rather the extremity and the Last Judgment of pseudophilosophy, of which signs were already seen in the stars, in the moon, and in the sun. I wait to hear sproutings of great things from the Peripetate to maintain the immutability of the skies, which I don't know how can be aved and covered up, since the sun itself indicates to us with most manifest sensible experiences. Hence I expect that the mountains of the moon will be converted into a joke and a pleasantry in comparison with the wisps of these clouds, vapors and smokings that are being continually produced, moved, and dissolved on the face of the sun."

Early in 1612, Galileo received the first of a series of letters from an anonymous astronomer who called himself Apelles and who claimed to have been the first to observe sunspots. Later it was established that Apelles was none other than the talented German astronomer Christoph Scheiner, but in time the competition

Above *This dramatic view of Jupiter's Great Red Spot and its surroundings was obtained by spacecraft Voyager in February 1979. Cloud details as small as 100 miles (160 kilometers) across can be seen here. The colorful, wavy cloud pattern to the left of the Red Spot is a region of extraordinarily complex and variable wave motion.*

between Scheiner and Galileo turned their relationship unpleasant.
It was also later realized that neither had been first to observe
sunspots, but that a Dutchman called Johann Fabricius (1587-1616)
saw them in June 1611.

When Galileo's *Treatise on Floating Bodies* was finally published
in the spring, it attracted both widespread acclaim and
considerable criticism from those out to get him. It was clear
that the ongoing hostility, although not yet serious, was a
concern for Galileo. He knew his sunspot observations were a
breakthrough; after all, the sun was not supposed to have any
imperfections, so after more examinations of the phenomenon
he decided to seek expert theological opinion, in order to
determine the exact theological stance that could be taken
against his studies. He asked Cardinal Carlo Conti (1585-1615),
the Prefect of the Inquisition in Rome, about the ramifications
of a changing heaven. Perhaps wisely, Conti refused to be
pinned down saying at one time that the Bible did not support
Aristotle's view, in fact, it implied the contrary, but also stating
that the Bible did not address the motion of the earth. Galileo
realized he had to tread carefully. Conti sent him a small book
that stated sunspots were small stars revolving around the sun.
This exemplified the Church's confused position. How could the
stars be unchanging yet also revolve around the sun?

Ever eager to make practical applications of his work, Galileo
proposed a method of using observations of Jupiter's satellites
to determine longitudes, especially at sea. He treated Jupiter as
a "clock" in the heavens and the satellites as its "hands." Sadly,
he never aroused much interest in his ingenious idea: It was
difficult to predict the motions of Jupiter's satellites, as these
sometimes appeared too soon and sometimes too late within
Galileo's "clock" concept. Interestingly, once one assumes
that the Earth and Jupiter both revolve around the sun those
problems are eliminated. Meanwhile, the painter Cigoli (1559-
1613) wrote to Galileo warning him that certain men who were
ill-disposed toward his notions had been meeting in a house
in Florence belonging to Archbishop Alessandro Marzimedici
(1563-1630), the head of the Episcopal See in Florence, to
discuss how best to attack those asserting that the Earth moved.

Above *Voyager 2 was the first spacecraft to observe the planet Neptune and its two satellites: Triton, the largest, and Nereid. The most obvious feature of the planet is its blue color, the result of methane in the atmosphere.*

At the end of 1612, after carefully examining the observational data he had acquired with his telescope, Galileo finally came out firmly in support of the Copernican system in the third of his letters on sunspots. In the missive, he also wrote about Saturn saying: "I tell you that this planet also, no less than horned Venus, agrees admirably with the great Copernican system on which propitious winds now universally are seen to blow to direct us with so bright a guide that little reason remains to fear shadows or crosswinds."

Curiously, during his observing Galileo had seen the light from a yet undiscovered planet, but had not realized what it was. On December 28, 1612 Galileo observed what he called a "star" east of Jupiter—this was in fact the planet Neptune which he noted 234 years before it was officially discovered.

The criticism of Galileo grew louder. As soon as Galileo published his letters on sunspots, an attack came from a Dominican friar and professor of ecclesiastical history in Florence, Father Niccolo' Lorini, also friend of Caccini. Preaching on All Soul's Day, Lorini said that Copernican doctrine violated Scripture, which clearly places Earth, and not the sun at the center of the universe. If Copernicus were right, he asked, what would be the sense of Joshua 10:13 stating, "So the sun stood still in the midst of heaven," or Isaiah 40:22 speaking of "the heavens stretched out as a curtain" above "the circle of the earth?"

Galileo was troubled by an incident involving his friend Benedetto Castelli (1578-1643), a mathematician and a Benedictine abbot. Castelli was in Pisa at the end of the year, and wrote to Galileo about what happened and the growing opposition to his works among the academics.

"On Thursday I dined at their Highnesses' table. The Grand Duke asked me how my lectures were attended. I entered into various minute particulars, with which he appeared much pleased. He asked whether I had a telescope. I answered that I had and with this I gave an account of my observation of the Medicean planets the preceding night."

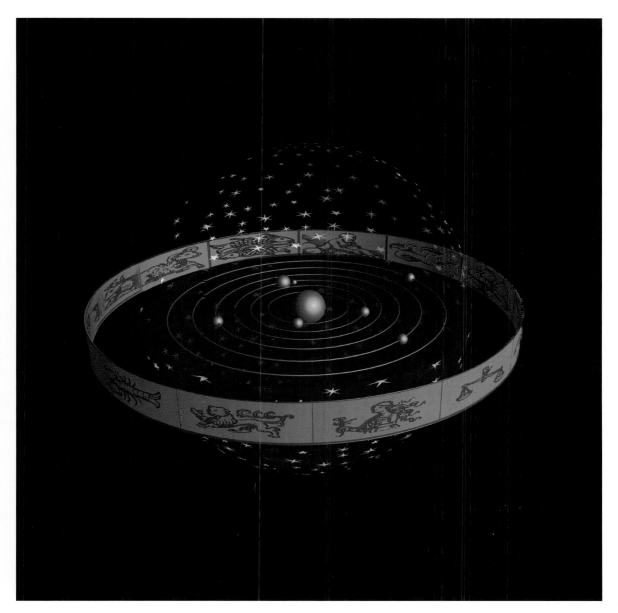

Above *The universe according to Copernicus shown with the astrological link. For those who studied the "heavens," astronomy and astrology were the same thing. They were also called The Celestial Sciences.*

Grand Duchess Christine (known as Madama Serenissima) took issue with Castelli. Others joined in.

"Hereupon some began to say that indeed these must be realities, and not deceptions of the instrument; and their Highnesses began to question Dr. Boscaglia [Cosimo Boscaglia, c. 1550-1621], the professor of physics, who answered that the existence of these planets could not be denied. I took occasion to add what I knew of your wonderful invention, and of your having fixed the periods of revolution of the said planets. Don Antonio was at table, who showed by his countenance how much pleased he felt with what I said. At length, after many solemn ceremonies, dinner came to an end, and I took leave; but scarcely had I quitted the palace when Madama Serenissima's porter came after me, and called me back. But before I narrate what followed, I ought to tell you that during dinner Boscaglia was talking privately to Madama for a while; and he said that, if it were conceded that the celestial novelties discovered by you were realities, then only the motion of the earth was incredible, and could not be, for the reason that Holy Scripture was manifestly contrary to it.

"I entered her Highness's apartment, where were the Grand Duke, Madama the Archduchess, Don Antonio, Don Paolo Giordano, and Dr. Boscaglia. Here Madama, after a few inquiries as to my condition in life, began to argue against me with the help of the Holy Scripture; and I, after making a proper protest, began a theological exposition in such a masterly manner that you would have been delighted to hear me. Don Antonio helped me, and so encouraged me that, though the majesty of their Highnesses was enough to appall me.

"The Grand Duke and the Archduchess were on my side, and Don Paolo Giordano brought forward a passage of Scripture very opportunely in my defense. So that at length Madama Serenissima was the only one who contradicted me, but it was in such a manner that I judged she only did it to draw me out, Signor Boscaglia said nothing either the one way or the other."

Galileo was worried. He did not wish to loose the support of the Grand Duchess. He took what, with hindsight, can be regarded as a significant step. He could not ignore the criticism, but wondered how far he should wade into theological waters. If he ventured too far he would encounter powerful and influential personalities and ultimately Rome itself. Some believe that it was at this point that Galileo made the biggest mistake of his life. A friend would later advise him to "write freely but be careful to keep outside of the sacristy." He rarely kept out of trouble, and his reply to Castelli played a significant part in the tumultuous events about to unfold (*right*).

What you said, as recounted to me...has given me occasion to consider again some general things concerning the carrying of Holy Scripture into disputes about physical conclusions, and some other particular things about the passage in Joshua proposed to you by the Grand Duchess Mother and the most serene Archduchess as contradicting the mobility of the Earth and the stability of the sun.

As to the first general question of Madame Christina, it seems to me that it was most prudently propounded to you by her, and conceded and established by you, that Holy Scripture could never lie or err, but that its decrees are of absolute and inviolable truth. I should only have added that although Scripture can indeed not err, nevertheless some of its interpreters and expositors may sometimes err in various ways, one of which may be very serious and quite frequent, that is, when they would base themselves always on the literal meaning of words. For in that way there would always appear to be in the Bible not only various contradictions, but even great heresies and blasphemies, since, literally it would be necessary to give to God feet and hands and eyes, and no less corporeal than human feelings, like wrath, regret and hatred, or sometimes even forgetfulness of things gone by and ignorance of the future.

Nature being inexorable and immutable, and caring nothing whether her hidden reasons and modes of operating are or not revealed to the capacities of men, she never transgresses the bounds of the laws imposed on her. Hence it appears that physical effects placed before our eyes by sensible experience, or concluded by necessary demonstrations, should not in any circumstances be called in doubt by passages in Scripture that verbally have a different semblance, since not everything in Scripture is linked to such severe obligations as is every physical effect.

And who wants to set bounds to the human mind? Who wants to assert that everything is known that can be known to the world?

I do not think it necessary to believe that the same God who has given us our senses, reason, and intelligence wished us to abandon their use, giving us by some other means the information that we could gain through them—and especially in matters of which only a minimal part, and in partial conclusions to be read in Scripture, for such is astronomy, of which there is in the Bible so small a part that not even the planets are named.

Galileo Galilei

Above *The Hubble telescope has proved that both Copernicus and Galileo were right about the immensity of the heavens.*

Galileo hoped that his letter to Castelli might bring about a reconciliation of faith and science, but it achieved precisely the opposite, increasing the tension between them. Either intentionally or perhaps without considering the consequences, Castelli made copies of this letter, and some found their way into the hands of Galileo's opponents. His enemies now had the ammunition they had been wanting all along. He had ventured onto their territory and they accused him of attacking Scripture and meddling in theological affairs. Galileo's destiny was beginning to unfold dangerously, although it took many months for the storm to gather force.

On December 21, 1614, at the church of Santa Maria Novella in Florence, Tommaso Caccini gave another incendiary sermon. However, it was not Caccini but his friend Niccolo' Lorini, who first put forward a complaint before the Inquisition. Lorini was from the same convent as Caccini, and a ringleader for local intrigues against Galileo. Once an accusation was sent to the Inquisition, it had to be investigated in secret. Caccini was called to Rome and questioned by the Inquisition, who had set about gathering all the gossip and innuendo concerning Galileo. Giorgio de Santillana, author of *The Crime of Galileo*, wrote of Caccini's testimony: "The whole deposition is such an interminable mass of twists and innuendoes and double talk that a summary does no justice to it."

Caccini said: "All our Fathers of this devout convent of St. Mark are of opinion that the letter contains many propositions which appear to be suspicious or presumptuous, as when it asserts that the language of Holy Scripture does not mean what it seems to mean; that in discussions about natural phenomena the last and lowest place ought to be given the authority of the sacred text; that its commentators have very often erred in their interpretation; that the Holy Scriptures should not be mixed up with anything except matters of religion. ...When, I say, I became aware of all of this, I made up my mind to acquaint your Lordship with the state of affairs, that you in your holy zeal for the Faith may, in conjunction with your illustrious colleagues, provide such remedies as may appear advisable. ...I, who hold that those who call themselves Galileists are orderly men and good Christians all, but a little over-wise and conceited in their opinions, declare that I am actuated by nothing in this business but zeal for the sacred cause."

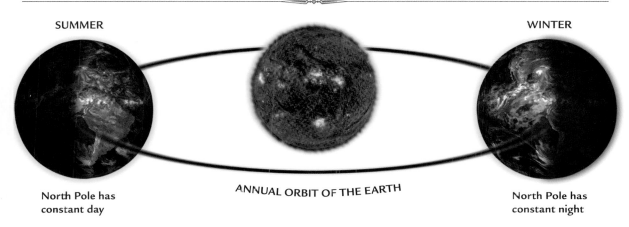

SUMMER

WINTER

North Pole has
constant day

ANNUAL ORBIT OF THE EARTH

North Pole has
constant night

ORBITING THE SUN

To anyone who stands and looks at the sky, it seems clear that the earth
stays in one place while everything in the sky rises and sets or goes around
once every day. Observing over a longer time, one sees more complicated
movements. The Sun makes a slower circle over the course of a year;
the planets have similar motions, but they sometimes turn around and
move in the reverse direction for a while in their retrograde motion.
Copernicus discussed the philosophical implications of his proposed
system, elaborated it in full geometrical detail, used selected astronomical
observations to derive the parameters of his model from a series of
astronomical observations, and wrote astronomical tables which enabled
one to compute the past and future positions of the stars and planets. In
doing so, Copernicus moved heliocentrism from philosophical speculation
to predictive geometrical astronomy.

Galileo's letter to Castelli was also considered part of the
evidence, but the copy of the letter that Lorini offered to the
Inquisition had been altered. Cardinal Mellini (dates uncertain),
president of the Congregation of the Index—a special congregation
that had been set up in 1571 to investigate writings that had
been denounced in Rome—ordered its secretary to write to the
Archbishop and Inquisitor of Pisa and quietly obtain and forward
to Rome the original letter of Galileo to Castelli. The Archbishop
sent for Castelli and requested the original, lying about his reasons
and telling him he was friends with the writer. Castelli had in the
meantime returned the letter to Galileo, but wrote him requesting
it back suspecting no subterfuge. Galileo, however, was suspicious
and delayed giving his friend an answer. Wishing to pave the way

diplomatically for Galileo's journey back to Rome, the Grand Duke of Tuscany wrote to friends and influential people at the Vatican:

"Galileo, a mathematician well known to your illustrious lordship, informs me that, having felt himself deeply aggrieved by the calumnies which have been spread by certain envious persons, to wit, that his writings contain erroneous opinions, he has of his own accord resolved to go to Rome, and has for this purpose asked my permission, having a mind to clear himself from such imputations."

Aware of the move against him, Galileo wrote to a friend, Monsignor Dini (dates uncertain), asking that his letters be forwarded to the influential Cardinal Bellarmine, the Church's chief theologian, and, if it could be arranged, Pope Paul V. Unfortunately for Galileo, the seventy-four-year-old Cardinal Bellarmine was no friend of novelties although, unlike some of Galileo's other detractors, he had at least looked through a telescope and given an audience to Galileo in 1611. In his innate conservatism, Cardinal Bellarmine saw the Copernican universe as threatening to the social order. To him and to much of the Church's upper echelon, the science of the matter was beyond their understanding—and in many cases their interest. They cared more for the administration and the preservation of Papal power than they did for getting astronomical facts right.

Bellarmine had another section of Scripture to quote (Psalms 19:1–6):
In the heavens the lord has set a tabernacle for the sun
Which is like a bridegroom coming out of his chamber
Rejoicing as a giant in running his course.
His going forth is from the high heaven,
And his circuit unto the ends of it;
And there is nothing hidden from the heart thereof.

"Let Galileo explain that," he offered.

In the summer of 1615, Galileo's health improved although his judgment did not. Feeling better, he considered visiting Rome. There exists evidence that the Grand Duke was not keen on such

Left *Atlas holding the world on his shoulders as portrayed in 1559. At the center are the elements: earth, water, air, and fire, after which come the celestial spheres in the order established in the second century by Ptolemy: Moon, Mercury, Venus, Sun, Mars, Jupiter, Saturn, and the fixed stars. Outside the fixed stars (the "firmament") are two additional spheres invented in the Middle Ages for theological reasons. Also shown in the band is the Zodiac, together with the celestial equator, the tropics of Cancer and Capricorn, and other circles.*

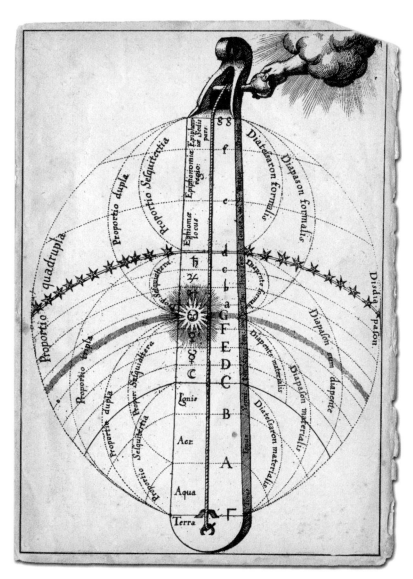

Above *Robert Fludd's seventeenth-century drawing of the universe as a monochord. Many shared Kepler's view of a harmonic universe.*

a visit. The Inquisition's interest in Galileo continued and was further piqued when they heard that a Father Ferdinando Ximenes (dates uncertain), Chancellor of Santa Maria Novella in Florence, had once told Caccini that he had listened to some of Galileo's followers saying that God is not substance, but accident. Further, the followers also stated that God is a sensitive being and that the miracles attributed to the saints are not true miracles. Subsequently, Galileo's reputation, profession, and even birthplace were the subject of inquiry.

The Grand Duke, Cosimo II, knowing of Galileo's desire to go to Rome wrote letters to members of the Medici court who were serving as cardinals in Rome: "Galileo, a mathematician well-known to your illustrious lordship, informs me that he feels himself deeply aggrieved by the calumnies which have been spread by certain jealous persons, to wit, that his writings contain erroneous opinions. He has asked, of his own accord to come to Rome and has asked my permission, for he has a mind to clear himself from such imputations. He wishes to speak the truth, to show his rectitude and pious intentions."

On December 3, 1615 Galileo set out for Rome. Ambassador Guicciardini (dates uncertain) was worried that Galileo's hot temper would get him into yet more trouble. He warned Galileo that it was not the place or the time to enter into disputes about the moon or to defend "novelties." Galileo did not listen. The Vatican believed that the new science was trespassing on religion. Galileo believed the opposite: that religion was trespassing on the domain of science. However, while visiting Rome he did not relive the triumphant experience he had had four years earlier. Things had changed: There was resentment against the ideas of Copernicus and also against Galileo personally.

Cardinal Bellarmine sat firmly at the center of the gathering storm. On the surface, Bellarmine was a man of unshakable belief in the literal meaning of the Bible. He had been the Church's chief weapon against heretic Protestants, and had sat in judgment over Giordano Bruno condemning him to the stake. At first sight he was doing God's work, but delving deeper his motives turned out not to be that straightforward.

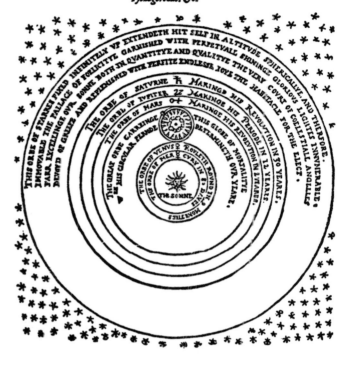

Above *Copernican system drawn by Thomas Diggs.*

Bellarmine stated his views on the Galileo controversy in a letter he wrote on April 12, 1615, to Father Foscarini (1565–1616), a highly respected monk from Naples. In the missive, Bellarmine indicated that Galileo could speak about the Copernican model "hypothetically, and not absolutely." Further, he wrote that "to affirm that the Sun, in its very truth, is at the center of the universe…is a very dangerous attitude and one calculated not only to arouse all scholastic philosophers and theologians but also to injure our faith by contradicting the Scriptures." Previously Bellarmine had written, "When there shall be a real demonstration that the sun stands still in the center of the universe, and that the Earth revolves around it, it will then be necessary to proceed with great caution in explaining those passages of Scripture which appear contrary to this."

With nineteen centuries of organized thought piling up to smother him," Galileo pleaded his case in a powerful summary of his thoughts on Scriptural interpretation and the evidence concerning the nature of the universe when he wrote to the

Medici Grand Duchess. He asked that his idea not be condemned "without understanding it, without hearing it, without even having seen it." The Duchess forwarded it to Rome where it sank without a trace.

Harmonia Mundi Sympathica, 10 Enneachordis totius naturæ Symphoniam exhibens.

Ennaeachor.on I	Enneach. II	Enneach. III	Enneach. IV	Enneach. V	Enneach. VI	Enneach. VII	Enneach. VIII	Enneach. IX	Enneach. X
Mundus Archetyp. DEVS	Mundus Sidereus Cœl.Emp.	Mundus Mineralis	Lapides	Planta	Arbores	Aquatilia	Volucria	Quadrupedia	Colores varij
Seraphim	Firmamentum	Salia,stellæ Minerales.	Aftrites	Herbæ & Flor.ftell.	Frutices Bacciferæ	Pifces ftellares	Gallina Pharaonis	Pardus	Diuerfi Colores
Cherubim	♄ Nete	Plumbum	Topazius	Helleborus	Cypreffus	Tynnus	Bubo	Afinus, Urfus	Fufcus
Troni	♃ Paranete	Æs	Amethiftus	Betonica	Citrus	Acipenfer	Aquila	Elephas	Rofeus
Dominationes	♂ Paramef.	Ferrum	Adamas	Abfynthifi	Quercus	Pfyphias	Falco Accipiter	Lupus	Flammeus
Virtutes	☼ Mefe	Aurum	Pyropus	Heliotropium	Lotus, Laurus	Delphinus	Gallus	Leo	Aureus
Poteftates	♀ Lichanos	Stannum	Beryllus	Satyrium	Myrtus	Truta	Cygnus Columba	Ceruus	Viridis
Principatus	☿ Parhypa.	Argentum Viuum	Achates Iafpis	Pæonia	Maluspunica	Caftor	Pfittacus	Canis	Cæruleus
Archangeli	☽ Hypate	Argentum	Selenites Cryftallus	Lunaria	Colutea	Oftrea	Anates Anferes	Ælurus	Candidus
Angeli	Ter.c ûEle. Proslamb.	Sulphur	Magnes	Gramina	Frutices	Anguilla	Stru thio camelus	Infecta	Niger

Above *The universe as a harmonious arrangement based on the number 9, from Athanasius Kircher's* Musurgia Universalis, Rome, 1650.

In Rome, few would listen to Galileo, and he was received cordially but kept at a distance. He wrote, "My business is far more difficult and takes much longer owing to untoward circumstances than the nature of it would require. I cannot communicate directly with those persons with whom I have to negotiate partly to avoid doing injury to any of my friends, and partly because they cannot communicate anything to me without running the risk of grave censure. So I am compelled with much pains and caution to seek out third persons who, without even knowing my object, may serve as mediators with the principals. ...I have also to set down some points in writing and to cause that they shall come privately into the hands of those I wish to see them."

Above *Tycho Brahe's planetary system, in a 1660 version that incorporates Galileo's discovery of four moons shown to its right. In the center is the Earth, encircled by the orbits of Mercury and Venus, and further out, those of Mars, Jupiter, and Saturn. An alternative position for the sun and its attendant satellites is shown in the lower half of the figure. Seated at the bottom right is Tycho Brahe, measuring a globe.*

Troublemaker Tommaso Caccini sought an interview with Galileo, supposedly to assure him that he had not been the prime mover in the dispute, and excusing his actions and sermons as having been commanded by his superiors. Galileo was unimpressed saying of him: "I perceived not only his great ignorance, but that he has a mind void of charity and full of venom."

Father Niccolo' Lorini made still more contemptible excuses, declaring that he knew nothing, and wanted to know nothing, of the merits of the pending controversy, and that he had only spoken in the first instance "for the sake of saying something," lest men should think that the Dominican fathers were asleep or dead. Both Lorini and Caccini knew not the power they had evoked. The Congregation of the Index, once set to the task of scenting out heresy, was not to be quieted until a victim could be sacrificed to it.

Galileo's position in Rome at this time is described in lively but somewhat exaggerated terms in the Medici ambassador's letter to the Grand Duke: "Galileo has chosen to follow his own opinion rather than that of his friends. Cardinal del Monte, and I as much as I could, besides many Cardinals of the Holy Office, have endeavored to persuade him to be quiet and make no more stir in this matter, but if he will hold this opinion, to hold it tacitly, without endeavoring to make others hold it too.

"For we feared that his coming here would be both

Below *Kepler's own drawing of his model of the five platonic solids.*

prejudicial and dangerous to him. But he, after importuning many of the Cardinals, threw himself on the favor of Cardinal Orsini [dates uncertain], and for this end procured from your Serene Highness a warm letter of recommendation to this Cardinal, who last Wednesday spoke in the Consistory to the Pope in favor of the said Galileo. The Pope answered that he would do well to persuade Galileo to give up this opinion.

"Orsini made some answer which roused the Pope to opposition. He cut the discussion short, saying he had referred the matter to the Congregation of the Holy Office. As soon as Orsini was gone, his Holiness had Cardinal Bellarmine called, and after discussion they decided that Galileo's opinion was erroneous and heretical."

When all the evidence against Galileo had finally been collected, a summary was forwarded to eleven theologians, called "Qualifiers." They were asked what they thought of the sun being at the center of the world and immovable of local motion, and the Earth not being at the center of the world, nor immovable, but moving according to the whole of itself, as well as possessing diurnal motion. Four days later, on February 23, 1616, the Qualifiers unanimously declared both propositions to be "foolish and absurd" and "formally heretical."

From this moment on, everything happened swiftly. Less than two weeks later, Pope Paul V—once described as "so averse to anything intellectual that everyone has to play dense and ignorant to gain his favor"—agreed with the conclusions. Then, according to the file of the Inquisition he "directed the Lord Cardinal Bellarmine to summon before him the said Galileo and admonish him to abandon the said opinion; and, in the case of his refusal to obey, the Commissary of the Holy Office is to enjoin him...to abstain altogether from teaching or defending this opinion and even from discussing it."

Galileo was summoned to appear before Bellarmine on February 25, 1616. Galileo was fifty-three years old, and appeared loud and unkempt. By contrast, Bellarmine, twenty years older, was small and reflective.

Bellarmine told Galileo that the views of Copernicus were wrong, and ordered him to abandon them. Also in the room was Inquisitor Cardinal Agostino Oreggi (1577-1635) who acted as witness. He later wrote that Galileo "remained silent with all his science and thus showed that no less praiseworthy than his mind was his pious disposition." The Inquisitor was more blunt than the Cardinal; Galileo was "not to hold, teach or defend it [Copernican theory] in any way whatever, either orally or in writing." If he did not obey, the Holy Office would bring formal charges against him of heresy.

Galileo must have remembered what had happened to Giordano Bruno at the hands of Bellarmine and he remained silent for a while and then, quietly acquiesced.

The Inquisition's account of the meeting went thus:

"At the palace, the usual residence of Lord Cardinal Bellarmine, the said Galileo, having been summoned and being present before the said Lord Cardinal, was...warned of the error of the aforesaid opinion and admonished to abandon it; and immediately thereafter...the said Galileo was by the said Commissary commanded and enjoined, in the name of His Holiness the Pope and the whole Congregation of the Holy Office, to relinquish altogether the said opinion that the sun is the center of the world and immovable and that the Earth moves; nor further to hold, teach, or defend it in any way whatsoever, verbally or in writing; otherwise proceedings would be taken against him by the Holy Office; which injunction the said Galileo acquiesced in and promised to obey."

When Galileo later complained of rumors to the effect that he had been forced to abjure and do penance, Bellarmine wrote out a letter denying the rumors, stating that Galileo had merely been notified of the decree and informed that, as a consequence of it, the Copernican doctrine could not be "defended or held."

Although Galileo stayed in Rome for many weeks following the injunction, it is clear that the Grand Duke, his Ambassador, and his advisors wanted Galileo to return home quickly fearing that he would get into more trouble if he remained. They wrote to him, "You have had enough of monkish persecutions and ought to know by this what the flavor of them is. His Highness fears that your longer stay in Rome would involve you in fresh difficulties and would therefore be glad if (as you have so far come honorably out of the affair) you would not tease the sleeping dogs any more,

Above *Title page from The* Dialogue Concerning the Two Chief World Systems. *Galileo's interlocutors are, left to right: Sagredo, Simplicio, and Salviati.*

and would return here as soon as possible. There are rumors flying about which we do not like, and the monks are all powerful."

Galileo's admonition stopped the Copernican movement dead in its tracks, and it also marked a period of silence for him. Back in Florence, he felt depressed and was at a loss as to what to do. His friends were worried about his health. Sagredo told him (*right*). However, Galileo could not leave science alone. He carried on observing, endured his rheumatism, enjoyed the attention of his daughter Suor Maria Celeste, took an interest in the ebbs and flows of the tides, and adjusted to a world that elevated mindless obedience over scientific understanding.

He did not give up his belief in what he had seen through his telescope and the Copernican interpretation of it. Many years later, under a different Pope, who claimed to be his friend, Galileo would face the gravest danger of his life when he stood before the very same Inquisition and was charged with violating the injunction.

"Philosophize comfortably in your bed and leave the stars alone."
Sagredo

CHAPTER SIX

HERESY

In 1622, Galileo wrote *The Assayer*, which was approved by the
Church and published in 1623. In 1624, he developed the first known
example of the microscope. In 1630, Galileo returned to Rome
to obtain a license to print the *Dialogue on the Great World Systems*,
published in Florence in 1632 amid an outbreak of the bubonic
plague. But in October of that year, he was ordered to appear
before the Holy Office in Rome.

In the years following the Inquisition's admonition of Galileo—that he should neither hold nor defend Copernican ideas—both he and the Church were lulled into what turned out to be a false sense of security. Both sides, because of lack of foresight and because of personalities, eventually stumbled toward an inevitable conflict. Initially, Galileo was relatively untroubled by the edict, which now seemed to him to have been a formality. Time and attitudes would change, he assumed. His belief was confirmed when the Vatican scrutinized the troublesome book by Copernicus and only required a few minor changes. The Earth-centered view of the cosmos was not in itself heretical as long as it was considered a mathematical fiction and not rigorously true, and investigating did not mean one strictly upheld it. It must have seemed to him that the danger was over. He knew he had to be somewhat careful when dealing with the Church, but he wrote, "I have always acted, and I shall continue to act, so as to shut the mouth of malice, and to show that a saint could not have shown more reverence for the Church nor greater zeal than I have done."

In the spring of 1617, Galileo took up residence in a spacious villa at Bellosguardo, meaning "beautiful view," over the valley of the river Arno toward Florence, close to his daughters Virginia and Livia, sixteen- and fifteen-years-old respectively. As both were illegitimate they could not have respectable marriages, and therefore both had become nuns at the Franciscan convent of San Matteo in nearby Arcetri. On taking their orders, their names

Below *The last villa at Arcetri where Galileo lived and died.*

Above *The view from Arcetri overlooking Florence*

changed to Suor Maria Celeste and Suor Arcangela. They lived a harsh, somewhat miserable, and poverty-stricken existence. Galileo was particularly fond of Maria Celeste and used to visit her often, relying on her increasingly as he got older.

The villa at Bellosguardo was pleasing. He took great delight in the garden and the night sky remained fascinating to observe with the telescope. Soon, however, Galileo felt isolated and frustrated, because he was unable to comment on the important subject of the nature of the universe and on his scientific observations. This frustration grew in time, and physically he was not well. His new friend Archduke Leopold of the Austrian Tyrol visited him in the spring of 1618, and found him in his bed with arthritis, muscle and kidney problems, and a hernia that made walking painful. Days, weeks, and even months would pass when Galileo would conduct his business exclusively from his bed. After Leopold's visit, Galileo wrote him saying he felt like a prisoner. Even when in the autumn of 1618 three comets appeared in the sky that excited the attention of every astronomer in Europe, Galileo's observations of the phenomenon were interrupted by his illness, and he was confined to his bed for almost their entire apparition. Later, Leopold was to write to him saying, "May God grant you a better health in the new year and everything that you could want in this world."

Meanwhile, at the Collegio Romano, Father Orazio Grassi (1583–1654), an architect, mathematician and fervent Jesuit, presented his thoughts on the recent appearance of the comets. Because they looked no larger when seen through a telescope, he reasoned that they must be distant, between the realms of the sun and the moon. Grassi said that this showed that the Copernican system was false.

Galileo's supporters wanted him to respond, but either due to ill health or perhaps to an untypical caution he did not. One of Galileo's students, Mario Guiducci (1585–1646), took it upon himself to answer Grassi, at first by himself and increasingly with the frail Galileo's help. When Guiducci eventually addressed the Collegio Romano with his opposing views to Grassi's statements, everyone knew that he was speaking with the voice of Galileo. In August 1619, Guiducci's studies were published in a pamphlet titled *Discourse on the Comets*, and although it appeared under his name, it was largely the work of Galileo himself. This could have been a purely scholarly publication, but Galileo, perhaps feeling safe in his bed from the implication his harsh words would have on others, could not resist making barbed comments on the original treatise by Grassi that would lead to trouble for him in the future. Grassi was offended and vowed to take revenge when the time was right.

Grassi quickly wrote a counter-reply in a book titled

Libra Astronomica (Astronomical Balance), as one of the comets had been observed passing through the constellation of Libra. He said he was using the metaphor of the balance to "weigh" Galileo's arguments, "all will be tested," he wrote and he found Galileo's philosophy wanting. In his tract, Grassi pulled no punches. "It is not safe for a pious man to assert the motion of the Earth," he wrote, along with, "he who is dutiful will call everyone away from Copernicus and will reject and spurn his recently rejected hypothesis."

Again, many friends urged Galileo to counterattack, especially fellow members of the Accademia dei Lincei in Rome, who, when Galileo once again proved reticent, began plotting an offensive of their own against Grassi. They approached Guiducci and asked him to write a reply. However, Guiducci had learned something from Galileo about writing letters and had become counterproductively aggressive, especially about Grassi. The respective positions became more entrenched and the resentment greater, although Galileo kept his silence on the matter. Within this backdrop, opinions hardened. Galileo had hoped the attitude would change, but when it did, it became more conservative still. Inevitably, the Inquisition formally banned Copernicus' book *De Revolutionibus* as well as Kepler's book, *Epitome on Copernican Astronomy* that supported it.

It was during this time that fate took a curious twist, and a new factor came into the simmering conflict. In 1621, Pope Paul V died. Galileo hoped for a more moderate pontiff to take his place, one who would see the virtues of the new science. The Cardinals at the Vatican came together for the election in the first fortnight of August, and the heat in Rome was suffocating. Many prelates and members of the Conclave fell ill and some even died. It was said that it was God's way of bringing them to a swift decision. Eventually Galileo's friend Maffeo Barberini was elected, and he took the name of Pope Urban VIII. As Cardinal, Urban had been on such intimate terms with Galileo as to sign himself his "affectionate brother." Indeed, he had even written Latin sonnets in praise of Galileo and his discoveries. Later the same year Cardinal Bellarmine died. Galileo was overjoyed: surely, the Church's attitude would now soften. He made plans to visit the new Pope and outlined a written response to Grassi's *Astronomical Balance*.

However, once again, Galileo made mistakes, and misinterpreted the situation, an error he would repeat several times in subsequent years. In his critique of Grassi's work, rather than arguing against Grassi's statements, he promoted himself as a philosopher. In his rebuke, he drifted from his obvious scientific expertise into theological matters. Some of Galileo's biographers say he fell into an ecclesiastical trap whereby his opponents could attack him on matters spiritual and not scientific. It is a fact that he fell into the trap slowly and probably without realizing it. His reply to Grassi grew larger and larger and soon he called his tome *The Assayer*.

When it was published, it caused a sensation in Rome, with its logic, wit, audacity, and its flirting with dangerous subjects. In fact, Galileo was wrong about the nature of comets—he believed they were tricks of light whereas Grassi believed they were real objects—but this was hardly the point. Ever the one to curry favor, Galileo dedicated the book to the new pope. The title page even showed Urban VIII employing a member of the Accademia dei Lincei Galileo was proud of, called Virginio Cesarini (1595-1624), for help and advice. Galileo wrote; "Philosophy is written in this grand book, I mean the universe, which stands continually open to our gaze, but it cannot be understood unless one first learns to comprehend the language and interpret the characters in which it is written. It is written

Left *Pope Urban VIII was a powerful pope. In this painting by Andrea Sacchi he celebrates Mass at the Church of Il Gesu'*

Right *Portrait of Galileo Galilei at his home in Arcetri.*

in the language of mathematics, and its characters are triangles, circles, and other geometrical figures, without which it is humanly impossible to understand a single word of it; without these, one is wandering around in a dark labyrinth."

The Assayer had one major outcome: it made the Jesuits, of which Orazio Grassi was a leading member, his eternal enemies. They would not forget. Not that the new pope himself liked the Jesuits: the book was read with delight at the dinner table by Urban VIII, who had written a poem lauding Galileo for his rhetorical performances.

However, in time, Pope Urban changed. It was as if being God's premier representative on Earth had gone to his head. He became progressively less tolerant and increasingly arrogant, and saw no need to moderate his whims or temper. He was no longer the promising young Cardinal who had observed Galileo's buoyancy experiments years before. He still seemed to hold some affection toward Galileo, but, over the years, the sentiment diminished.

Galileo's poor health delayed his visit to Rome to see the new pope until Easter of 1624. His friends at the Papal Court anxiously watched the new pope's temper and increasing paranoia coupled with an obsession with astrology. Despite this, Galileo and his friends had high hopes that the new science would flourish under Urban. Federico Cesi, the founder of the Accademia dei Lincei, wrote, "Under the auspices of the excellent, learned, and benignant Pontiff, science must flourish. Your arrival will be welcome to his Holiness. He asked me if you were coming, and when; and, in short, he seems to love and esteem you more than ever."

When he was well enough to travel to Rome, Galileo brought one of the first microscopes, a device he had been working on for years and which he had developed into a fine art. He hoped it would offer an entertaining scientific diversion. Many were astounded by the sights of fleas magnified as big as eggs and "horrendous things" seen floating in drops of fetid water. Galileo stayed in Rome for about two months. During this time, he had no less than six interviews with the pope, who, on his departure, presented him with "a fine painting, two medals, one of gold and the other of silver, and a promise of a pension of sixty crowns for his son." It seems that Urban was anxious to appear as Galileo's chief patron. Pope Urban even took the opportunity to write the young Grand

Duke Ferdinand a letter on Galileo's return to Florence, stating he was a person worthy of protection and favor. The pope's letter read, "...we find in him not only literary distinction, but also the love of piety, and he is strong in those qualities by which pontifical good-will is easily obtained. Now, when he has been brought to this city to congratulate us on our elevation, we have very lovingly embraced him, nor can we suffer him to return to the country whither your liberality calls him without an ample provision of pontifical love. And that you may know how dear he is to us, we have willed to give him this honorable testimonial of virtue and piety. And we further signify that every benefit which you shall confer upon him, imitating or even surpassing your father's liberality, will conduce to our gratification."

It is no wonder Galileo felt that he had an eternal friend in the pope. However, in private Urban exhibited unstable and tyrannical tendencies, refusing to take advice from others for he believed he was infallible. He was known to become hysterical toward any who approached the outer limits of criticism. For days, he was locked away with his astrologers.

Emboldened by the pope's friendliness, Galileo tentatively expressed his support for Copernicus publicly in a pamphlet that took the form of a letter of reply to Francesco Ingoli (1578–1649), a priest who some years before had written a treatise on the Copernican system. Galileo distributed a few copies of his pamphlet among his friends, and sent one to Monsignor Giovanni Ciampoli (1590–1643), the pope's secretary, asking him to choose a fitting time to show it to the pontiff. Ciampoli informed Galileo in a letter written on December 28, 1625 that the pope was greatly pleased with both the manner and the matter of it.

Galileo's relationship with his daughter Suor Maria Celeste had grown closer over the years. She wrote him a letter at this time that provides an insight into her nature. "Of the preserved citron you ordered, I have only been able to do a small quantity. I feared the citrons were too shriveled for preserving, and so it has proved. I send two baked pears for these days of vigil. But as the greatest treat of all I send you a rose, which ought to please you extremely, seeing what a rarity it is at this season. And with the rose you must accept its thorns, which represent the bitter Passion of our Lord, while the green leaves represent the hope we may entertain that through the same Sacred Passion we, having passed through the darkness of this short winter of our mortal life, may attain to the brightness and felicity of an eternal spring in heaven, which may our gracious God grant us through his mercy. Here I must stop. Sister Arcangela joins me in affectionate salutations; we should both be glad to know how you are at present."

Galileo had been considering for some time writing a significant book that drew together all he knew about science and the Copernican system. Initially it was to be only about his theory of the tides, which he believed proved that the Earth was in motion, but it soon expanded into much more. If Galileo hoped that it would be published without incident, with the accommodating pope's tacit approval, he was very mistaken.

In the book, Galileo presented his arguments about the new science as revealed through his telescope and the writings of Copernicus as a discussion between characters holding different viewpoints who conduct a conversation spanning four days. This was a literary form often used at the time. The characters included two philosophers and a layman, modeled after his friends and enemies, a ploy that was transparent to all readers who knew him. One of the characters, based on Galileo's friend Filippo Salviati, argued the Copernican position and presented Galileo's views directly. His other friend Sagredo provided both the name and the persona for the intelligent layman who was at first neutral in the story. The final character Simplicio was portrayed as a dedicated follower of Ptolemy and Aristotle, and he presented the Church's view against the Copernican position. He was probably modeled on Galileo's enemies Cremonini and delle Colombe. The character's name refers to the sixth-century philosopher Simplicius (c. 490–c.560 AD), but the Inquisition would object to the name's meaning of "simpleton" which was, of course, deliberate on Galileo's part.

In the spring of 1627, Suor Maria Celeste fell ill with a fever. Bad food at the convent was to blame. She wrote to her father asking for a little money to enable her to procure such comforts as were necessary to aid her recovery. She said that the convent bread was very bad; the wine was sour, and the beef coarse and uneatable. She pleaded for a tough old hen in the poultry yard at

Bellosguardo, from which she could make some weak broth. Since the death of their mother, Giulia, in 1620, the brothers Galilei had communicated little, if at all. Galileo's position at the court of the Grand Duke of Tuscany was secure, and in his brother Michelangelo's estimation, it was good enough to secure him riches as well. Michelangelo reckoned that Galileo might keep the honor for himself: the riches, his brother considered, ought to be divided freely with the members of the family who had been less fortunate, or less enterprising. Michelangelo had been to Germany, but the country was unstable and disturbed, and he considered it unlikely that he would ever make a fortune in Munich. Michelangelo's thoughts turned to more serious matters than his lute, and he contacted his brother on the possibility of returning to Florence.

Galileo had at one point considered employing Massimiliana, the sister of his sister-in-law Clara Galilei, as his housekeeper. Massimiliana lived in Michelangelo's house, attending to the children and the cooking, and he could not spare her. Michelangelo curiously offered to send his own wife, Clara, to act as his brother's housekeeper. "This arrangement," he wrote, "would be well and faithfully governed, and I should be partly lightened of an expense which I do not know how to meet; for Clara would take some of the children with her, who would be an amusement to you and a comfort to her. I do not suppose that you would feel the expense of one or two mouths more. At any rate, they will not cost you more than those you have about you now, who are not so near akin, and probably not so much in need of help as I am."

Galileo seemed perpetually on the verge of serious illness, and he had a relapse that caused his daughter the greatest anxiety. "Only in one respect does cloister life weigh heavily on me; that is, that it prevents my attending on you personally, which would be my desire were it permitted. My thoughts are always with you, and I long to have news of you daily. As you were not able to see the steward the day before yesterday, I send him again today with these two pieces of preserved citron as an excuse. You will be able to tell him if there is anything we can do for you, and if the quince was to your liking, because, if so, I might prepare another for you. I, Sister Arcangela, and our friends here, pray the Lord without ceasing for your restoration to health."

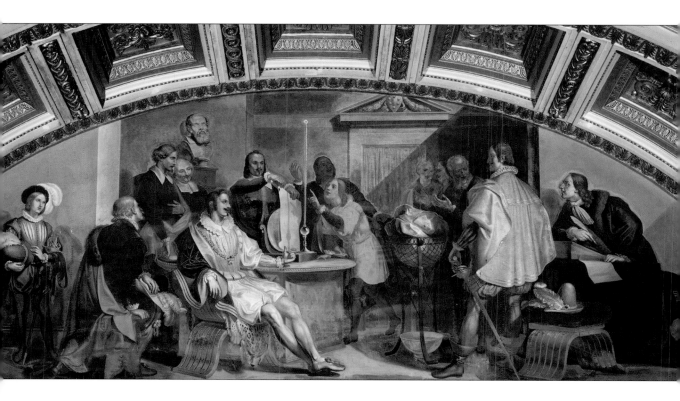

In July 1628, after six years at Pisa University, where he studied law and mathematics, Galileo's son, Vincenzo, then twenty-two years old, took his doctor's degree, which Galileo himself, more than thirty years before, had refrained from taking on account of the unavoidable expenses it involved. It was Galileo's wish to see his son employed in some branch of the civil service, but Vincenzo preferred living an idle life at home under the pretense of aiding his father in his experiments. It was another drain on his expenses.

At the beginning of November 1628, Suor Maria Celeste wrote again in great distress. Her sister's health had been worse than usual, even though ill health was the normal condition of most nuns at San Matteo's convent. Certain symptoms had appeared which were both unusual and alarming. Maria Celeste knew her sister ought to be bled, and had sent for the surgeon. However, she did not have enough money to pay him, nor to procure the necessary comforts for the poor invalid, "who had fallen into her usual melancholy mood, for the love of God" and asked her father to send Vincenzo, if the weather allowed, so that she could tell him their troubles.

During this time, as he worked on his great book intermittently, Galileo's own health was so poor as to put a stop to his visits at the convent. Such strength as he had he devoted to his scientific labors. His daughter wrote again: "You may think from my long silence that I had forgotten you, just as I might suspect that you had forgotten the road to our abode, from the length of time which has elapsed since you came that way. However, as I know that the reason of my silence is that I have not a single hour at present that I can call my own, so I think of you, that not forgetfulness, but press of business, keeps you from coming to see us. It is some comfort to have Vincenzo's visits, as by this means we get news of you that we can rely on. The only thing, which I am sorry to hear of, is that you are in the habit of going into the garden of a morning. I cannot tell you how grieved I am to hear this, for I feel sure that you are rendering yourself liable to just such another lingering illness as you had last winter. Do pray leave off this habit of going out, for your own sake and of your daughters who desire to see you arrive at extreme old age; which will not be the case unless you take more care of yourself than you do at present. As far as my experience goes, if ever I stand still in the open air without some covering on my head, I am sure to suffer for it. And how much more hurtful must it be for you!"

Galileo had also been mending the convent clock: "Pray your lordship pardon me if I am tedious; my love for you carries me sometimes beyond bounds. I did not sit down to write about my own feelings, but to tell you that if you could manage to send back the clock on Saturday evening, the sacristan, whose duty it is to call us to matins, would feel much obliged. But if you have not been able to set it to rights yet, never mind; for it will be better for us to wait a little longer than to have it back before it has been properly put in order."

"I want to make you some rosemary conserve, but I must wait till you send back some of my glass jars, because I have nothing to put it in. At the same time, if you have any empty jars or phials which are in the way, I should be glad to have them for the pharmacy."

Such were the privations at the convent that the abbess

attempted suicide. Maria Celeste told Galileo; "Now that the tempest of our many troubles is somewhat abated, I will no longer delay telling you all about them; hoping thereby to lighten the burden on my own mind, and desiring also to excuse myself for writing twice in such a hurry, and not with the respect I owe you. The fact was, that I was half out madness of my senses with fear (and so were the other nuns) at the furious behavior of our mistress—who during these last few days has twice endeavored to kill herself. The first time she knocked her head against the floor with such violence that her features became quite monstrous and deformed. The second time she gave herself thirteen wounds, of which two were in the throat. You may imagine our consternation on finding all over her blood and wounded in this manner. But the strangest thing of all was, that at the time that she inflicted these injuries upon herself, she made a noise to attract somebody to her cell, and then she asked for the confessor. In confession, she gave up to him the instrument with which she had cut herself, in order that nobody might see it (though, as far as we can guess, it was a penknife). It seems that, though mad, she is cunning. And we must conclude that this is some dark judgment upon her from God, who lets her live when according to human judgment she ought to die, her wounds being all dangerous in the surgeon's opinion. In consequence, she has been watched day and night. At present we are all well, thank God, and she is tied down in her bed, but has just the same frenzy as ever, so that we are in constant terror of something dangerous happening."

Galileo's great work, the *Dialogue on the Ptolemaic and Copernican Systems*, was finally concluded in the beginning of March 1630. As a mark of the affection he felt for his pupil and patron, and so that the work might appear under the most favorable auspices, it was dedicated to the Grand Duke Ferdinand of Tuscany.

In the book, the first of the four-day conversation between the protagonists focused on the Aristotelian concept of the perfect sphere, with heaven made of perfection in all things from its composition to its motions, and in which only perfect circular motions were allowed. Galileo also suggested that there might be life in other worlds. This was dangerous ground, as it was one of the heretical notions that had led Giordano Bruno to be burnt

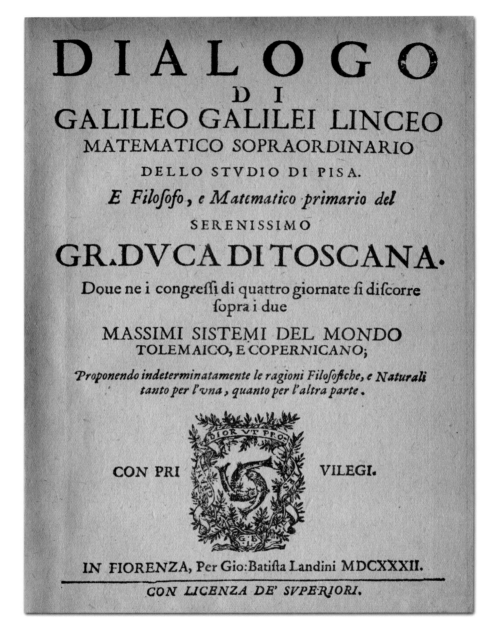

Above *Title page of the* Dialogue *by Galileo, 1632.*

at the stake in 1600. However Galileo felt confident, as Cardinal Bellarmine, Bruno's prosecutor and Galileo's judge, had been dead for years. He then compounded his impious suggestions by saying that it was not for the scientist to comment whether Christ had visited these worlds and redeemed the creatures he found there. In the *Dialogue* the characters converse as follows:

SIMPLICIO:

On earth I continually see herbs, plants, animals generating
and decaying; winds, rains, tempests, storms arising; in a word,
the appearance of the earth undergoing perpetual change.
None of these changes are to be discerned in celestial bodies,
whose positions and configurations correspond exactly with
everything men remember, without the generation of anything
new there or the corruption of anything old.

SALVIATI:

But if you have to content yourself with these visible, or rather
these seen experiences, you must consider China and America
celestial bodies, since you surely have never seen in them these
alterations that you see in Italy. Therefore, in your sense, they
must be inalterable.

SIMPLICIO:

Even if I have never seen such alterations in those places with
my own senses, there are reliable accounts of them; besides
which... those counties being a pan of the earth like ours, they
must be alterable like this.

SALVIATI:

But why have you not observed this, instead of reducing
yourself to having to believe the tales of others? Why not see it
with your own eyes?

SIMPLICIO:

Because those countries are far from being exposed to view,
they are so distant that our sight could not discover such
alterations in them.

SALVIATI:

Now see for yourself how you have inadvertently revealed
the fallacy of your argument. You say that alterations, which
may be seen near at hand on Earth, cannot be seen in America
because of the great distance. Well, so much the less could they
be seen in the moon, which is many hundreds of times more
distant. And if you believe in alterations in Mexico on the basis
of news from there, what reports do you have from the moon to
convince you that there are no alterations there? From your not
seeing alterations in heaven (where if any occurred you would
not be able to see them by reason of the distance, and from
whence no news is to be had), you cannot deduce that there are
none, in the same way as from seeing and recognizing them on
earth you correctly deduce that they do exist here.

SIMPLICIO:

Among the changes that have taken place on earth I can find some so great that if they had occurred on the moon they could yen well have been observed here below. From the oldest records we have it that formerly, at the Straits of Gibraltar, Abila and Calpe were joined together with some lesser mountains which held the ocean in check; but these mountains being separated by some cause, the opening admitted the sea, which flooded in so as to form the Mediterranean. When we consider the immensity of this, and the difference in appearance which must have been made in the water and land seen from afar, there is no doubt that such a change could easily have been seen by anyone then on the moon. Just so would the inhabitants of earth have discovered any such alteration in the moon; yet there is no history of such a thing being seen. Hence there remains no basis for saying that anything in the heavenly bodies is alterable, etc.

SALVIATI:

I do not make bold to say that such great changes have taken place in the moon, but neither am I sure that they could not have happened. Such a mutation could be represented to us only by some variation between the lighter and the darker parts of the moon, and I doubt whether we have had observant selenographers on earth who have for any considerable number of years provided us with such exact selenography as would make us reasonably conclude that no such change has come about in the face of the moon. Of the moon's appearance, I find no more exact description than that some say it represents a human face; others, that it is like the muzzle of a lion; still others, that it is Cain with a bundle of thorns on his back. So to say, "Heaven is inalterable, because neither in the moon nor in other celestial bodies are such alterations seen as are discovered upon the earth has no power to prove anything."

CON PRI VILEGI.

On the second day, the characters in the book considered the ideas of Copernicus directly as well as the heretical concept of the Earth's motion. On day three, they considered the suggested motion of the Earth around the sun. On the final day, Galileo presented his cherished theory on the sea tides that he believed irrefutable evidence of the motion of the Earth.

Neither Galileo's fame nor the Grand Duke of Tuscany's protection were sufficient to warrant a good publication of the book, given its subject matter and the author's reputation. Before the book could be printed, it had to have the approval of the Church, and, in order to obtain this with as little delay as possible, Galileo was advised to go to Rome and deal with the Vatican censors in person. He was optimistic. Monsignor Niccolo' Riccardi (1585–1639), Master of the Sacred Palace, and the person responsible for granting publication licenses at the Vatican, had given his word that, as far as he was concerned, he should meet with no difficulty in obtaining the desired license: "The Barberini family [the pope's own family whose members included powerful members of the Vatican's curia] and the Cardinals are well disposed," he told Galileo. Moreover, Pope Urban had expressed his regret at Galileo's 1616 admonition and had declared that, had it depended on him, that decree would not have been published.

Suor Maria Celeste, hearing how deeply her loving father was immersed in study and of the strain a journey to Rome would be for him, wrote to say she did not wish him to shorten his precious life for the sake of immortal fame. She urged him to take care of himself for his children's sake, and reminded him tenderly that though they all loved him, she loved him with a love far surpassing that of the others, and that his suffering was affliction and torment to her.

Galileo arrived in Rome in the beginning of May 1630, with no reason to be dissatisfied with his reception at the Papal Court. Cardinal Francesco Barberini (1597–1679), the pope's nephew, had declared that he had no better friend than the pope; and a person who had ventured to bring him a vile and mendacious report concerning Galileo's private life, was met with a stern rebuff. Galileo's meetings with the Vatican's censors went well, and only minor adjustments to his text were requested. The pope did not

Above *Kepler's view of the universe linked the planets with the platonic solids and their cosmic geometries. Mars as dodecahedron, Venus as icosahedron, Earth as sphere, Jupiter as tetrahedron, Mercury as octahedron, Saturn as cube.*

object to the publication of the *Dialogue* if two conditions were met. First, the title should show plainly that the Copernican system was treated as mere hypothesis. Secondly, the book itself was to conclude with an argument of his own, which Pope Urban considered unanswerable by others given his elevated stature. Rather than forego the publication Galileo consented. He felt that the minds capable of following his reasoning in favor of the Copernican system, would be no more convinced of the falsity of it by the Pope's argument than he was himself.

This was the time when the plague was rife all over Europe; Galileo did not want to remain long in Rome, and left with a provisional approval that well satisfied him. However, behind the scenes, the pope, who was paranoid and pompous, was not happy, and his mind began to turn. Galileo returned to Florence and waited: as weeks passed he started to worry, and there were ominous signs that not all was well.

A letter arrived from Father Benedetto Castelli (1578–1643), a friend and long-time supporter, saying that for reasons "which he dared not commit to paper," Galileo should publish his book in Florence and not Rome as soon as possible. Prince Federico Cesi, the founder of the Accademia dei Lincei, who had offered to supervise the printing of the book, died suddenly in Rome. Galileo had lost the best, most influential, and the most enlightened friend he possessed out of Tuscany.

The plague broke out with such virulence that for a time communication between Rome and Florence became impossible. Galileo, anxious for the book to be published quickly, tried

to obtain permission from the papal authorities to have it printed in Florence. Eventually he received approval, however it was granted by the Inquisitor-General and the Vicar-General of Florence, not by Rome directly. To oversee the printing he traveled almost daily into Florence from Bellosguardo, an arduous journey for his age and infirmity.

Above *Galileo, portrayed in this painting in his old age, had many students and followers.*

The printing of Galileo's most important book took place amid social catastrophe. The plague, already rife within the city gates, now began to spread to the suburbs, and thousands died. Even Bellosguardo, the fashionable suburb, was not spared; one of Galileo's own household, a glass-blower, was taken. Vincenzo, seized by panic, fled with his wife to the nearby city of Prato, leaving his father alone and his child with a nurse in the neighborhood at Bellosguardo. Suor Maria Celeste wrote to her father: "I am troubled beyond measure at the thought of your distress and consternation at the sudden death of your poor glass-worker. I entreat you to omit no possible precaution against the present danger."

The plague continued in Florence, and no potions, prayers, or spells could stop it. The city's board of health ordered a quarantine enjoining even neighbors from communicating with each other. This only served to exacerbate the panic, as, unable to talk, no one knew who might be dead or dying, and consequently rumors were rife. Galileo paid little attention. He had his beloved tower, his telescope, his own thoughts and his daughter's almost daily letters. She told him that she had kept all his missives, and read them during such moments as she could snatch from her many duties. In turn, Galileo waited for her letters, and found time to write to her and keep her assured of his own well-being and that of his small

Below *A detail from the Madonna of Mercy, showing the devastation wrought by the bubonic plague.*

grandson. He also sent her money and gave medical advice, besides presents such as the Lent food that his daughters liked best. In one letter Suor Maria Celeste, after referring to her brother Vincenzo's behavior as being "the fruit of this ungrateful world," states: "I am quite confused at hearing that you keep my letters; I fear that your great love for me makes you think them more perfect than they are. But let it be as you will; if you are satisfied, that is enough for me."

Galileo waited and waited for final approval of his book to come directly from Rome. In the spring of 1631, he was still waiting and he grew increasingly impatient and frustrated, when matters took another worrying turn. The Vatican's chief licenser, Niccolo' Riccardi, had been in further discussions with Pope Urban about the book and concluded that he had given permission for the book to be printed without full consideration. He knew he had committed an error and hastily tried to remedy it. Despite the decision of the Inquisitor in Florence, Niccolo' Riccardi told Galileo he wished to look at the book a third time. Galileo, anxious and uneasy at this further delay, eventually wrote to Andrea Cioli (1573–1650), the Grand Duke's State Secretary (*right*).

"As your illustrious lordship knows, I went to Rome for the purpose of getting permission to publish my Dialogues, and to this end I put them in the hands of the most Reverend Father the Master of the Sacred Palace...that he might look at them with the most particular attention, and note if there were any doubtful matter, or any conceit of imagination which required correction: which, at my own request also, he did most thoroughly. And as I entreated the Reverend Master to give the required license, and affix his name thereto, his Reverence signified his wish to read the whole book through once more. This was done, after which he returned me the book, with the permission signed with his own hand. Whereupon I, having been at Rome for two months, returned to Florence, intending to send back the book (as soon as I had written the index, the dedication, and a few other necessary things) to the most illustrious and eminent Prince Cesi, head of the Lyncean Academy, who had always intended the printing of my other works. But, owing to the death of this Prince and the interruption of communication, I was hindered from printing my work at Rome, and decided on having it done here. I had found and arranged matters with an able printer and publisher, and procured the permission of the Reverend Vicar and of the Inquisitor. ... I informed the Reverend Master of the Palace of all that had taken place, and of the impediments in my way touching the printing of my book at Rome. Whereupon he sent to tell me through my lord Ambassador (Niccolini), that he wished to have another look at the book, and that I was to send him a copy. ...Weeks and months ago I heard from Father Benedetto Castelli, that he had often met the Reverend Master, who had given him to understand that he was going to send back the preface and the end, arranged to his entire satisfaction. But this has not been the case, and I hear no word of its being sent back: the book has been thrown aside into some corner, and my life is wasting away, and I am in continual trouble."

Galileo Galilei

In the autumn of 1631, Galileo moved to a new villa to be even closer to his daughters. Eventually, the papal censors decided to write the first and last parts of the book. To Galileo their text was nonsensical and he had it printed in different type to distinguish it from "his" words, and set the text as coming from the character Simplicio. Finally, his great work, the *Dialogue*, was published in January 1632. The book began with the following sentence: "Dialogue by Galileo Galilei, Mathematician Extraordinary of the University of Pisa, and Principal Mathematician and Philosopher of the Most Serene Grand Duke of Tuscany, in which, in a conference lasting four days, are discussed the two principal systems of the world, proposing indeterminately the philosophical arguments on each side."

At first, the ecclesiastical authorities were pleased with the book, but slowly the dissenting voices of Galileo's enemies who saw their long-awaited opportunity to attack him became louder. The story is told that that his old adversary Christoph Scheiner first saw a copy in a bookshop in Rome, and that he shook and turned red with anger as he read it. As he left, he told the owner of the bookshop that it was outrageous that he had to pay gold to purchase the book only to be slandered and humiliated. Scheiner complained loudly and his cries were heard by none other than another of Galileo's foes, Father Orazio Grassi.

Then, Pope Urban himself turned against Galileo. Father Riccardi heard of this and decided to take action. At the beginning of August, he ordered the Florentine officials to sequester every copy of the book in booksellers' shops, not only in the States of the Church, but throughout Italy. Landino (dates uncertain), Galileo's publisher, received an injunction to suspend the publication of the book, and to forward to Rome all copies in his possession. Landino answered that he had not a single copy left. Alarmed and worried, the Grand Duke ordered Francesco Niccolini, his ambassador in Rome, to demand an explanation of the Pope's sudden caprice:

"I have orders to signify to your Excellency his Highness's exceeding astonishment that a book, placed by the author himself in the hands of the supreme authority in Rome, read and read again there most attentively, and in which everything, not only with the consent but at the request of the author, was amended,

Above *The five platonic solids that Kepler believed to be the building blocks of the universe. The sphere contains all of them (as shown in the reflected crystal).*

altered, added, or removed at the will of his superiors, which was here again subjected to examination, agreeably to orders from Rome, and which finally was licensed both here and there, and here printed and published, should now become an object of suspicion at the end of two years, and the author and printer be prohibited from publishing any more."

The letter further stated that, considering the manner in which Galileo had handled his subject, the Grand Duke strongly suspected that this suspension of the book was not so much caused by zeal for purity of doctrine, as by dislike to the person of the author. He therefore desired that the reasons of the suspension be set forth clearly in writing, and forwarded to Galileo, who felt strong in the consciousness of his own innocence, and had declared this only a fresh instance of his enemies' malignity. Moreover, he had offered to leave Tuscany and forfeit the Grand Duke's favor, if the charge of holding heretical doctrine could be fairly proven against him.

In September, Francesco Niccolini was granted an audience with Pope Urban in which he was to present the letter from the Medici Grand Duke. The meeting was disastrous for all concerned.

At first, Niccolini thought that the meeting was going well but suddenly, without any warning, the pope exploded with anger after Galileo's name was mentioned. He shouted that Galileo was meddling in matters that were dangerous for him. Niccolini, by now in an embarrassing and difficult position, said that Galileo's book had received Vatican approval. Urban interjected, shouting that he had been deceived. Niccolini did the best he could, saying that Galileo should be told of these new objections, but Urban, once Galileo's friend, replied with anger that it was the practice of the Holy Office to arrive at censure and then tell the defendant to recant. Niccolini's astonishment at the pope's sudden change of sentiments was such that "he thought the world must be falling to pieces." He decided not to show the pope the Grand Duke's letter of protest.

Urban was furious at the fact that Galileo had placed his argument, which the pope had insisted be part of the book and unanswerable because of his elevated stature, in the mouth of Simplicio. The pontiff desired Niccolini to tell the arrogant Duke that in a matter such as this, he ought to bring the offender to

punishment, instead of protecting him, and that the Duke had best not meddle in this business because he would not come out of it with honor.

Urban passed the matter to the Inquisition, and his orders initiated a formal and unstoppable procedure. The Inquisitors were all grave and saintly men who, after weighing every word in the book and with due deliberation, would reach the pope's own conclusions. Pope Urban now maintained that the book contained the most perverse matter that could come into a reader's hands. The pope made it understood that he had acted with extraordinary kindness toward Galileo in not sending his book at once to the Inquisition, and that Galileo ill deserved this leniency, for he had dared to deceive him. "I found the pope greatly incensed, and indeed full of ill-will to our poor Signor Galilei," Niccolini reported to the Grand Duke, "so your lordship may think in what a state of mind I returned home yesterday morning."

As soon as the matter was handed over to the Inquisition, it became a Church secret. Galileo's friends rushed to tell him what had happened. Among these was Fillipo Magalotti (1558–?), whose letter to Mario Guiducci stated:

"On Monday morning, I was in the Church of San Giovanni, when the most reverend Father Riccardi, having heard that I was there, came to seek me. He signified to me that it would be agreeable to him were I to give up the whole of the copies of Signor Galileo's book of Dialogues which I had brought from Florence, promising to return them in ten days at the farthest. I answered that I regretted infinitely not being able to comply with his wish, for of the six copies which I had brought, five were for presentation, and his Reverence knew that they had already been presented; that is to say, one to his Eminence Cardinal Barberino, one to himself, one to the Ambassador of Tuscany: and the other three, one to Monsignor Serristori, a member of the Congregation of the Holy Office; one to Father Leon Santi, a Jesuit; and lastly, one for myself. I told him that it was impossible to ask to have back again those copies from the persons to whom they had been presented; and as for my copy, it was in the hands of Signor Girolamo Reti, the Prefect's Chamberlain, and I was not sure but what his Excellency himself was reading it. He must know that in this particular it was impossible for me to satisfy him. At the very utmost Riccardi confided to Niccolini, that he had got himself into serious trouble in consequence of his having given Galileo the permission to print; that there were a few phrases in the book which certainly wanted alteration, and that the end was not at all in accordance with the beginning; besides which, the pages containing the license to publish were printed in a totally different type to the rest of the book."

Niccolo' Riccardi, who had first given permission for the book to be printed on behalf of the pope, was heard to state that the less said in defense of Galileo the better. He also told Ambassador Niccolini, under the seal of secrecy, that he had found in the books of the Holy Office that sixteen years before Galileo had been absolutely forbidden to hold or discuss the opinions that he had presented in the Dialogue.

In early October, the Inquisitor of Florence, in the presence of the Apostolic Proto-Notary and two monks as witnesses, informed Galileo that he was being investigated by the Holy Inquisition in Rome. Meekly Galileo said he was a good Catholic, promised obedience, and signed the order. which then received the signatures of the Proto-Notary and witnesses. Shortly after, he wrote an anguished letter to Cardinal Francesco Barberini, in which he begged for a delay in going to Rome, and sent the letter to Ambassador Niccolini for delivery. Niccolini refused to present it, saying that it would only find its way to the Inquisition as Francesco Barberini was the pope's nephew. Niccolini solicited, unasked, the protection for Galileo of two other cardinals, Marzio Ginetti (1586–1671), the pope's intimate friend, and Boccabella (dates uncertain), Assessor of the Congregation. "To these," he wrote, "I represented his advanced age [he said Galileo was seventy-five years old, whereas in fact he was sixty-eight], his weak health, and the danger of traveling at this time, besides the discomfort attendant on performing quarantine. But as these are men who hear and answer not, this morning I spoke to his Holiness on the subject, placing strongly before him Signor Galileo's prompt submission, and begging him to take compassion on the poor man, so very

Right *Galileo at the tribunal of the Holy Office being condemned of heresy.*

aged, and whom I so love and revere."

However, Niccolini's plea was to no avail. Pope Urban declared it was absolutely necessary for Galileo to appear in person before the Inquisition. Niccolini suggested that so extreme was the old man's weakness that if his Holiness insisted on his coming, he would, in all probability, fall ill and die on his way to Rome. The pope would not change his mind, and said that he might perform the journey as slowly as he pleased, but that come he must. He insisted Galileo had brought all this trouble on himself.

On November 9, Galileo was again called before the Inquisition of Florence. He declared himself ready and willing to obey; and said that his only reason for delaying his journey to Rome had been his age and infirmities, which were at that moment of a serious nature, requiring medical treatment. The pope granted a delay of one month. On December 8, the Inquisitor of Florence, wrote that his vicar had seen Galileo, who was confined to his bed, and declared that he was utterly incapable of undertaking a journey until his condition would be somewhat ameliorated. The pope and the Congregation chose to treat this statement as a mere subterfuge. His friends feared that Galileo would be thrown into prison. Ambassador Niccolini advised Galileo to get a medical certificate as proof of his illnesses, but the angry pope would not be deterred. The certificate was received in Rome with shakes of the head.

Eventually, because of the mounting pressure, even the Grand Duke was convinced that Galileo would only damage his own cause by further delay. Therefore, Galileo set out on his weary journey in one of the grand-ducal litters on January 20, 1633.

It was a very different journey from the one that he had undertaken eight years before in very different circumstances. Prince Federico Cesi, his supporter and counselor, was dead. His health, never good, was now a constant source of concern, and lately his eyesight had begun to fail. Not only was the time of year most unpropitious, as the tramontana, the seasonal wind of Tuscany, was at its most biting and persistent during January and February, but the region through which he was traveling was bleak and inhospitable, and its inhabitants dangerous. Near Ponte Centino, where he was forced to stay for over two weeks, he found the countryside infested with thieves. He had written to his daughter, notwithstanding his dim eyesight. In

her answer, Suor Maria Celeste, after expressing grief at her father's forced delay in such a wretched habitation, deprived of every comfort, entreated him "to keep up his spirits, and put his whole trust in God, who never forsakes those who trust in Him."

When Galileo finally arrived in Rome, he was received as the honored guest of the Tuscan Ambassador. The following day, he paid a visit to the Ex-Assessor of the Holy Office, Monsignor Boccabella, who also received him with great kindness, and gave him some advice as to which behavior would be most prudent to adopt under the circumstances. His advice, in essence, was that Galileo should talk to no one. Feeling buoyant at Boccabella's sympathy, Galileo went on to pay the required official visit to the new Assessor of the Inquisition and its Commissary. The Commissary was not at home; but a certain Girolamo Matti, a great friend of the Commissary, was in, and Ambassador Niccolini thought that a visit to him might be useful, particularly as Signor Girolamo Matti had expressed his admiration for Galileo.

"For this day," Niccolini wrote, "it was impossible to do more. Tomorrow I shall endeavor to see Cardinal Barberini, in order to recommend Galileo to his protection. I shall also ask his Eminence to intercede with the pope for him to be allowed to stay in this house, instead of sending him to the Holy Office, in respect to his age, his reputation, and his readiness in obeying the mandate."

Two days after Niccolini wrote again, saying that as far as could be gathered, the Tribunal of the Holy Inquisition did not seem inclined to act with severity. In the meantime, Francesco Barberini, the pope's nephew, had sent a message asking that he not visit him and until he had further notice, remain within doors.

Monsignor Serristori (dates uncertain), one of the counselors of the Inquisition, and an old friend of Galileo, came twice under pretense of paying him a visit, but as he talked all the time of nothing but the trial, it was soon discovered that he had been sent by the Holy Office to discover Galileo's private opinions. "I think," Niccolini wrote, "that I have succeeded somewhat in cheering up the good old man, by what I have told him of the steps being taken in his favor. Yet he constantly expresses his wonder at this persecution." Eventually news came through that the charges against Galileo had been reduced to one—disobedience. At this Galileo was pleased because he felt that he could

"On that one charge I shall be able to clear myself without much trouble when the grounds of my defense are known."
Galileo

easily counter such an accusation (*left*).

In late February, Ambassador Niccolini had another audience with the pope hoping that he would find him in better temper. He did not, and, if anything, the pope was even more angry with Galileo. Niccolini knew that the main charge related to Galileo's oath in 1616 was that he would not discuss the question of the Earth's motion. Niccolini hoped that Galileo would get off on a technicality as the injunction had actually said that Copernican views were not to be held or defended, but he knew that the chances were slim. The pope's fury was undiminished. He believed that he was being lampooned as Simplicio in the book, and told Niccolini that Galileo was lucky not to have been clapped in irons and thrown into the dungeons. The wheels of the Inquisition continued to turn unbeknownst to all. Galileo heard nothing for weeks. It was a deliberate ploy on the part of the Inquisition: they wanted him to suffer.

CHAPTER SEVEN

BUT IT DOES MOVE

In February 1633, Galileo arrived in Rome to undergo trial for heresy
by the Holy Inquisition. He was frightened and unwell and Pope
Urban VIII was merciless toward his old friend. The court issued
a sentence of condemnation and forced Galileo to abjure publicly.
He was confined in Siena and eventually, in December 1633, he was
allowed to retire to his villa in Arcetri under house imprisonment.
In 1634, he was deprived of the support of his beloved daughter,
Suor Maria Celeste, who died prematurely. In 1638, when he was
almost totally blind, *Discorsi e Dimostrazioni intorno a Due Nuove
Scienze* (Discourses and Demonstrations on Two New Sciences) was
published in Leiden, The Netherlands, where there was no pope
or Church that condemned scientific treatises.

"May God forgive Galileo for meddling with these subjects," the pope was heard to say as Galileo waited to be called before the Inquisition. Ambassador Niccolini knew that Galileo was being watched and his demeanor was reported to the pope. He advised him to say nothing and be completely compliant. Niccolini had lived in Rome long enough to know that in terms of rumor and intrigue it was like no other place in the world. From the very beginning, he suspected that, in spite of the apparent leniency shown to Galileo, the Holy Office might have severity in mind, especially if the unpredictable pope wanted to make an example of him. Its present kindness might be a stratagem to make Galileo incriminate himself. In a dispatch dated February 27, Niccolini wrote of an audience with the pope in which he had notified the pontiff of Galileo's arrival. Without doubt, a fact the pope already knew.

During the audience, Pope Urban told Niccolini that Galileo had once been his friend and he was sorry to displease him now, but he must do "what was best for the furtherance of the Christian faith." Niccolini replied that he thought Galileo's opinion only went as far as God being omnipotent, and it was as easy to him to make the world go round as not. At this, however the pope flew into a rage. Certain passages in the *Dialogue* seemed to prove the pope's point; there were portions of Galileo's text that were frightening to the holy father because their implications opened the door to a new philosophy that asked painful questions of his

Below *A painting depicting the procedures of Galileo's trial.*

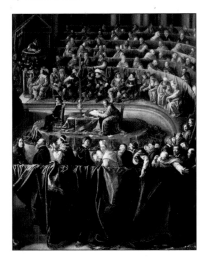

Above *Galileo was forced to go to Rome and await his trial by the Holy Inquisition.*

faith, questions for which the Church had no answer. Predictably, perhaps, the pope recoiled from what was unfamiliar and what made him feel powerless. Some of these difficult tracts of text were focused between Simplicio, the character in the book meant to represent both the pope and the entrenched views of the Church, Salviati who supported the Copernican system, and Sagredo who maintained neutrality of views:

SIMPLICIO:

It is not to be denied that the heavens may surpass in bigness the capacity of our imaginations, nor that God might have created them a thousand times larger than they really are; but we ought not to admit anything to be created in vain, or useless, in the universe. Now we see this beautiful arrangement of the planets, disposed round the earth at distances proportioned to the effects they are to produce on us for our benefit. To what purpose, then, should a vast vacancy be afterwards interposed between the orbit of Saturn and the starry spheres, containing not a single star, and altogether useless and unprofitable? To what end? For whose use and advantage?

SALVIATI:

Methinks we arrogate too much to ourselves, Simplicio, when we assume that the care of us alone is the adequate and sufficient work and limit beyond which the Divine wisdom and power does, and disposes of, nothing. I feel confident that nothing is omitted by God's providence which concerns the government of human affairs; but that there may not be other things in the universe dependent upon his supreme wisdom, I cannot, with what power of reasoning as I have, bring myself to believe. So that when I am told of the uselessness of an immense space interposed between the orbits of the planets and the fixed stars, empty and valueless, I reply that there is temerity in attempting by feeble reason to judge the works of God, and in calling vain and superfluous every part of the universe which is of no use to us.

SAGREDO:

Say rather that we have no means of knowing what is of use to us. I hold it to be one of the greatest pieces of arrogance and folly that can be in this world, to say, because I know not of what use Jupiter and Saturn are to me, that therefore these planets are superfluous; nay, more, that there are no such bodies in existence. To understand what effect is worked upon us by this or that heavenly body (since you will have it that all their use must have a reference to us) it would be necessary to remove it for a while, and then the effect which I find no longer produced in me I may say depended upon that star. Besides, who will dare say that the space that they call too vast and useless, between Saturn and the fixed stars, is void of other bodies belonging to the universe? Must it be so because we do not see them? Then in that case the four Medicean planets and the companions of Saturn came into the heavens when we began to see them, and not before! And, by the same rule, the innumerable host of fixed stars did not exist before men saw them. The nebulae, which the telescope shows us to be constellations of bright and beautiful stars, were, until the telescope was discovered, only white flakes. O, presumptuous! Nay, rather, O, rash ignorance of man!

Galileo's friends rallied around him during the tense period of waiting. Niccolini's wife, Caterina Riccardi, wrote to Galileo's daughter Suor Maria Celeste, begging her to persuade her father to behave precisely as if he were in his own house. However, Maria Celeste did not possess a true appreciation of her father's situation. Before Galileo was called for his first examination in front of the Inquisition, she wrote: "I want you to bring me a present on your return, which I trust is not far off. I am sure that at Rome copies of good pictures are easily obtained and I should like you to bring me a little picture the size of the enclosed piece of paper." Without doubt, contact with home left Galileo feeling sad.

The first examination of Galileo was scheduled for April 12, at the Palace of the Holy Office. As a special mark of pontifical favor, Niccolini had been told beforehand. Because Galileo was suffering from a severe attack of gout, Niccolini asked the pope to postpone the day, but this request was denied, and the Ambassador was told that Galileo's presence was absolutely necessary. In fact, it was completely unprecedented for a defendant to be at liberty while in course of examination before the Holy Office, a rule that ensured the complete secrecy of the proceedings. Usually the suspect was thrown into a dirt-floored dungeon, even if belonging to the nobility. Brave as ever, Niccolini suggested that surely the purpose of the Inquisition would be served if Galileo were forbidden to talk about his examinations. However, the pontiff was immovable, and Niccolini wisely desisted.

Niccolini implored Galileo that in order to bring the matter to an end as soon as possible, he should submit to anything the Holy Inquisition chose as a verdict, even if he held and believed the doctrine of the Earth's motion. Niccolini's advice was to confess, get out of the Vatican, and return home. However, the Ambassador's practical advice afflicted Galileo severely, "ever since yesterday has Galileo been in such a state of prostration that I have my fears for his life. I shall beg that a servant may be allowed him, and as much comfort as the place will admit of. Meanwhile we are all doing our best to console him."

The first examination took place before Commissary-General Carlo Sincere (dates uncertain), and Cardinal Vincenzo Maculani (1578–1667), and only lasted a few minutes. The questions were dry

Left *Trials by the Holy Inquisition were rampant in Catholic Europe. This painting by Francisco Goya shows three condemned men forced to wear hats during their trial to show stupidity and error.*

and simple, and meant to set a trap for Galileo. He replied simply with the occasional taunt that the cardinals ignored, knowing only too well they had all the time they wanted to interrogate Galileo and that this was merely the beginning. Galileo was asked whether he knew the reason of his being cited before the tribunal. He answered positively and was then, to his astonishment, returned to custody in the Holy Office, not in the prison, but in three rooms in the building where the interrogation was taking place. From Niccolini's account, it appears that Galileo was received with great courtesy by the Commissary-General, who had him installed in the apartments. Galileo was told he was at liberty to take the air in the court. His own servant was permitted to be with him, and his meals were taken to him twice a day from Niccolini's house. The Commissary declared that Galileo owed this gentle treatment in great part to the goodwill of Cardinal Barberini, the nephew of Pope Urban, who had been untiring in his efforts to mollify the pope's resentment. Tormented with the gout, deprived of the stimulating and comforting company of the Ambassador and his gracious wife, Galileo tolerated his imprisonment with a degree of impatience at variance with his natural serenity. Inwardly he was angry and frightened, but he resolved not to show it.

Suor Maria Celeste wrote to Galileo on April 20, 1633: "I have just been informed of your being imprisoned in the Holy Office. This, consoling though on the one hand, it grieves me much. Only be of good cheer. Do not let yourself give way to grief, for fear of the effect it would have on your health. Turn your thoughts to God, and put your trust in Him, who, like a loving Father, never forsakes those who trust in Him unceasingly. My dearest lord and father, I have written instantly on learning this news of you, that you might know how I sympathize with you in your distress. Perhaps when you know this, it will not be quite so hard to bear. I have mentioned what I have just heard to no creature in this house, choosing to make my joy and gladness common to all, but to keep my troubles to myself. Consequently, everybody is looking forward joyfully to seeing you back again. And who knows? Perhaps even while I am writing, the crisis may be past, and you may be relieved of all anxiety. May it be the Lord's will in whose keeping I leave you."

Above *The ultimate punishment for crimes of heresy was burning at the stake, a public act that was witnessed by crowds in public squares all over Catholic Europe.*

Nothing was achieved during Galileo's first interrogation, and Galileo's copy of the note from Cardinal Bellarmine in 1616 about not supporting Copernican views helped his case as it showed that the copy the Inquisition had of the note was imperfect. The second examination took place on April 30. Acting on Niccolini's advice, Galileo, on being invited to speak, tried to remove, if possible, from the minds of his judges the impression that he had written with willful malice against the Church on the subject of the Earth's motion. His error, he confessed, had been a vain ambition, and pure ignorance and inadvertence, but he declared himself innocent of willful disobedience. At the end of the second interrogation, he requested permission to speak. He suggested, as proof that he was not holding the forbidden doctrine, to add to his book one or two more dialogues in order to balance the arguments in favor of the Copernican theory contained in the body of the book. This did not seem to impress the Inquisitors who, once again, thought they had made little progress, and decided it was time for a change of approach: from then on, they would be kind and then harsh. Galileo was released and sent back to Niccolini's house that very evening, to the astonishment and delight of the whole household. Not less was the rejoicing at his daughters' San Matteo's convent on hearing the good news.

Cardinal Maculani visited Galileo at his residence and the two began to talk informally. Maculani told Galileo bluntly that the pope would not be convinced by any scientific observations and then he alluded to the very thing Galileo had been expecting and fearing most of all. Torture was a common procedure in the trial of suspected heretics, so common in fact that it was no disgrace to the torturer to carry it out. He told Galileo, subtly, but in no uncertain terms, that he was in danger of being strapped to the rack. He was not shown the instruments of torture, but Galileo could imagine them, and if torture did not elicit the required result, there was always the stake Maculani told him. The Cardinal also told him that days before three heretics had been put to the flames in the Campo dei Fiori.

This was too much for Galileo to bear: the fear, anguish, and tension broke Galileo and he collapsed into a mass of self-doubt. Had he gone too far in his book? He was now in doubt as to

Above *Galileo had been full of confidence earlier in life and found joy in scientific experimentation. However, his old age was made miserable and lonely by his public condemnation.*

whether he had, after all, followed the correct procedure. Maculani reported that the formal interrogations were closing and that, given Galileo's comments, the trial was nearing its end.

On May 10, Galileo was called before the Inquisition for the third time, and informed that eight days had been assigned to him for defense. He felt he did not need the time. This was not the same Galileo who had broken down before Maculani. He had thought of another way of presenting his actions, and made the mistake of trying to be too clever. He presented Cardinal Bellarmine's certificate, dated May 26, 1616. From the context of this certificate, he said that he believed himself at liberty to write as he had written. His defense was a repetition of the answers and explanations that he had given during his earlier examinations, and he ended it with a touching appeal to the mercy of the court.

Curiously, several of the Inquisitors—the two sitting at the large oak table who questioned Galileo, and the eight others who sat watching—were inclined to be lenient with him, because many of them actually favored the Copernican theory, and nearly all liked Galileo personally. More than one squirmed in their chair at the great scientist's humiliation and discomfort. However, Galileo had not said in public what he had said in private to Maculani, and the trial would be worse for him because of this. Galileo had been given to understand that the trial would terminate favorably, he would receive a lenient sentence from a kind and forgiving pope, and that he would shortly be able to return to Florence and resume his old pursuits. Unbeknownst to Galileo, this would now not be allowed to happen: Galileo had to be silenced.

A decree was issued by Pope Urban on June 16, that ordered a further examination for June 21, in which Galileo would be pressed again about his motives in writing the Dialogue. He was to be threatened with torture. If the menace had no effect, he was to be made to renounce the work and its views, and be imprisoned for as long a time as would please the Congregation. He was also to be told to abstain from discussing the Copernican theory in any way whatever, either for or against, under pain of being treated as a relapsed heretic, which meant the stake. Galileo could take no more. He said (*left*)

"I am in your hands; do as you please with me."
Galileo

Galileo capitulated and crumpled. Pope Urban obtained what he had wanted, and to Ambassador Niccolini it appeared that the pope's anger had suddenly melted in sorrow. The pontiff protested that he would have willingly treated Galileo with greater leniency due to his regard for the Grand Duke, but that Galileo's views, being contrary to Holy Scripture, had to be prohibited whatever the personal cost to him. Additionally, it would be necessary to inflict some salutary chastisement on him for having transgressed the decree of 1616. Others who might follow Galileo's line of reasoning had to be deterred. The pope declared that, though it was impossible to give Galileo a free pardon, he wished to afflict him as little as possible. He might be confined to a monastery for a while. "Of the personal punishment I have as yet said nothing to Galileo," Niccolini wrote, "in order not to distress him by telling him everything at once; also by his Holiness' orders, who did not wish to add to his troubles; and also because there may be a change of opinion." However there was none.

Above *Pope Urban VIII was an arrogant man who took personal revenge on Galileo. In this tapestry he is depicted concluding peace in Papal States.*

The judgment against Galileo read as follows:

By the mercy of God, Cardinals of the Holy Roman Church, Inquisitors of the Holy Apostolic See, in the whole Christian Republic specially deputed against heretical depravity:

It being the case that thou, Galileo, son of the late Vincenzo Galilei, a Florentine, now aged seventy, wast denounced in this Holy Office in 1615:

That thou heldest as true the false doctrine taught by many, that the sun was the centre of the universe and immovable, and that the Earth moved, and had also a diurnal motion. That on this same matter thou didst hold a correspondence with certain German mathematicians: That thou hadst caused to be printed certain letters entitled *On Solar Spots*, in the which thou didst explain the said doctrine to be true. And that, to the objections put forth to thee at various times, based on and drawn from Holy Scripture, thou didst answer, commenting upon and explaining the said Scripture after thy own fashion: And thereupon following was presented (to this tribunal) a copy of a writing in form of a letter, which was said to have been written by thee to such a one, at one time thy disciple, in which, following the position of Copernicus, are contained various propositions contrary to the true sense and authority of the Holy Scripture:

This Holy Tribunal desiring to obviate the disorder and mischief which had resulted from this, and which was constantly increasing to the prejudice of the Holy Faith; by order of our Lord (Pope) and of the most Eminent Lords Cardinals of this supreme and universal Inquisition, the two propositions of the stability of the sun and of the motion of the Earth were by the qualified theologians thus adjudged:

That the sun is the centre of the universe and doth not move from his place is a proposition absurd and false in philosophy, and formally heretical; being expressly contrary to Holy Writ. That the Earth is not the centre of the universe nor immovable, but that it moves, even with a diurnal motion, is likewise a proposition absurd and false in philosophy, and considered in theology ad minus erroneous in faith.

But being willing at that time to proceed with leniency towards thee, it was decreed in the Sacred, Congregation held before Our Lord (Pope) on February 25, 1616, that the most Eminent Lord Cardinal Bellarmine should order thee, that thou shouldst entirely leave and reject the said doctrine; and thou refusing to do this, that the Commissary of the Holy Office should admonish thee to abandon the said doctrine, and that thou wast neither to teach it to others, nor to hold or defend it, to which precept, if thou didst not give heed, thou wast to be imprisoned: and in execution of the said decree, the following day in the palace and in the presence of the said most Eminent Lord Cardinal Bellarmine, after having been advised and admonished benignantly by the said Lord Cardinal, thou didst receive a precept from the then Father Commissary of the Holy Office in the presence of a notary and witnesses, that thou shouldst entirely abandon the said false opinion, and for the future neither uphold nor teach it in any manner whatever, either orally or in writing: and having promised obedience, thou wast dismissed.

And to the end that this pernicious doctrine might be rooted out and prevented from spreading, to the grave prejudice of Catholic truth, a decree was issued by the Sacred Congregation of the Index, prohibiting books which treated of the said doctrine, which was declared to be false and entirely contrary to Holy Scripture.

And there having lately appeared here a book printed in Florence this past year, whose superscription showeth thyself to be the author, the title being: Dialogue of Galileo Galilei on the *Two Great Systems of the World, the Ptolemaic and the Copernican:* and the Sacred Congregation having been informed that in consequence of the said book, the false opinion of the mobility of the Earth and the stability of the sun was daily gaining ground; the said book was diligently examined, and was found openly to transgress the precept which had been made to thee, for that thou in the said book hadst defended the said already condemned opinion, which had been declared false before thy face: whereas thou in the said book by means of various subterfuges dost endeavor to persuade thyself that thou dost leave it undecided and merely probable. The which however is a most grave error, since in no way can an opinion be probable which has been declared and defined to be contrary to Holy Scripture.

Wherefore by Our order thou wast cited before this Holy Office, in which being examined upon oath, thou didst acknowledge thyself to have written and caused to be printed the said book. Thou didst confess that, ten or twelve years previously, after having received the precept above mentioned, thou didst begin to write the said book; that thou didst ask for a license to print it, without signifying to those from whom thou didst receive such license, that thou hadst a precept forbidding thee to hold, defend, or teach in any way whatever such doctrine.

Thou didst likewise confess, that the said book is in more places than one so written that the reader might form an idea that the arguments brought forward in favor of the false opinion were pronounced in such guise that by their efficacy they were more apt to convince than easy to be overturned; excusing thyself for having fallen into an error so alien, sayest thou, to thy intention, for that thou hadst written in form of a dialogue, and for the natural complacence with which each one doth view his own subtlety in showing himself more acute than the common herd of men in finding even for false propositions ingenious discourse to make them apparently probable.

And a convenient period having been assigned thee for thy defense, thou didst produce a certificate written by the hand of the most Eminent Lord Cardinal Bellarmine, procured by thee, as thou saidst, for the purpose of defending thyself from the calumnies of thy enemies, who had said that thou hadst abjured and hadst been punished by the Holy Office. In the which certificate it is written that thou hadst not abjured, neither hadst thou been subjected to punishment, but that only the declaration made by Our Lord (Pope) and published by the Sacred Congregation of the Index had been made known to thee, the which contains that the doctrine of the Earth's motion and of the stability of the sun is contrary to Holy Scripture, and may therefore neither be defended nor held: and that whereas in the said certificate no mention was made of two particulars of the precept, to wit, ... it was to be thought that in the course of fourteen or sixteen years thou hadst lost all remembrance of it ; and that for this same reason thou hadst been silent respecting the precept when thou didst ask for license to print the said book. And all this thou saidst not to excuse thy error, but that it might be attributed to a vain ambition rather than to malicious intent. But from the said precept produced in thy defense, thou hast aggravated thy fault; whereas, the said opinion being therein declared contrary to Holy Writ, thou hast nevertheless dared to treat of it, to defend it, and to persuade that it was probable; nor doth justify thee the license which thou didst extort with craft and cunning, not having notified the precept which had been given thee.

And, as it appeared to Us that thou hadst not said the whole truth concerning thy intention, We judged it to be necessary to proceed to the rigorous examination of thee, in which (without prejudice to any of the things confessed by thee, or deduced against thee, as above, respecting thy said intention) thou answeredst like a good Catholic. Therefore, having seen and maturely considered the merits of thy case, with thy above mentioned confessions and excuses, We have adjudged against thee the herein-written definite sentence.

Invoking then the Most Holy Name of Our Lord Jesus Christ, and
of His most glorious Mother Mary, ever Virgin, for this Our definite
sentence, the which sitting pro tribunal, by the counsel and opinion of the
Reverend Masters of theology and doctors of both laws, Our Counselors,
we present in these writings, in the cause and causes currently before Us,
between the magnificent Carlo Sinceri, doctor of both laws, procurator
fiscal of this Holy Office on the one part, and thou Galileo Galilei, guilty,
here present, confessed and judged, on the other part:

We say, pronounce, sentence, and declare, that thou, the said Galileo,
by the things deduced during this trial, and by thee confessed as above,
hast rendered thyself vehemently suspected of heresy by this Holy Office,
that is, of having believed and held a doctrine which is false, and contrary
to the Holy Scriptures, to wit: that the sun is the centre of the universe,
and that it does not move from east to west, and that the Earth moves
and is not the centre of the universe: and that an opinion may be held and
defended as probable after having been declared and defined as contrary
to Holy Scripture; and in consequence thou hast incurred all the censures
and penalties of the Sacred Canons, and other Decrees both general and
particular, against such offenders imposed and promulgated. From the
which We are content that thou shouldst be absolved, if, first of all, with a
sincere heart and unfeigned faith, thou dost before Us abjure, curse, and
detest the above-mentioned errors and heresies, and any other error and
heresy contrary to the Catholic and Apostolic Roman Church, after the
manner that We shall require of thee.

And to the end that this thy grave error and transgression remain not
entirely unpunished, and that thou mayst be more cautious for the future,
and an example to others to abstain from and avoid similar offenses,
We order that by a public edict the book of *Dialogues of Galileo Galilei*
be prohibited, and We condemn thee to the prison of this Holy Office
during Our will and pleasure ; and as a salutary penance We enjoin on thee
that for the space of three years thou shalt recite once a week the Seven
Penitential Psalms, reserving to Ourselves the faculty of moderating,
changing, or taking from, all or part of the above mentioned pains and
penalties.

And thus We say, pronounce, declare, order, condemn, and reserve in this
and in any other better way and form which by right We can and ought.

Of the ten judges, three refused to sign the decree. However, after the judgment was decreed against Galileo, it was now time to humiliate him publicly.

On June 22, Galileo was led to the great hall of the Inquisition at Santa Maria Sopra Minerva. There, before the supreme magistracy of the Holy See, the Pope alone being absent, Galileo was made to kneel and hear the sentence, which declared him "vehemently suspected of heresy," and condemned him to imprisonment during the pleasure of the Holy Office. As a salutary penance, he was ordered to say the Penitential Psalms once a week for three years. He was then made to recite the abjuration dictated to him beforehand by the vengeful pope (left).

I, Galileo Galilei, son of the late Vincenzo Galilei of Florence, aged seventy years, tried personally by this court, and kneeling before You, the most Eminent and Reverend Lords Cardinals, Inquisitors-General throughout the Christian Republic against heretical depravity, having before my eyes the Most Holy Gospels, and laying on them my own hands; I swear that I have always believed, I believe now, and with God's help I will in future believe all which the Holy Catholic and Apostolic Church doth hold, preach, and teach. But since I, after having been admonished by this Holy Office entirely to abandon the false opinion that the sun was the centre of the universe and immovable, and that the Earth was not the centre of the same and that it moved, and that I was neither to hold, defend, nor teach in any manner whatever, either orally or in writing, the said false doctrine; and after having received a notification that the said doctrine is contrary to Holy Writ, I did write and cause to be printed a book in which I treat of the said already condemned doctrine, and bring forward arguments of much efficacy in its favor, without arriving at any solution : I have been judged vehemently suspected of heresy, that is, of having held and believed that the sun is the centre of the universe and immovable, and that the Earth is not the centre of the same, and that it does move.

Nevertheless, wishing to remove from the minds of your Eminences and all faithful Christians this vehement suspicion reasonably conceived against me, I abjure with a sincere heart and unfeigned faith, I curse and detest the said errors and heresies, and generally all and every error and sect contrary to the Holy Catholic Church. And I swear that for the future I will neither say nor assert in speaking or writing such things as may bring upon me similar suspicion; and if I know any heretic, or one suspected of heresy, I will denounce him to this Holy Office, or to the Inquisitor and Ordinary of the place in which I may be. I also swear and promise to adopt and observe entirely all the penances that have been or may be by this Holy Office imposed on me. And if I contravene any of these said promises, protests, or oaths, (which God forbid!) I submit myself to all the pains and penalties that by the Sacred Canons and other Decrees general and particular are against such offenders imposed and promulgated. So help me God and the Holy Gospels, which I touch with my own hands. I, Galileo Galilei aforesaid, have abjured, sworn, and promised, and hold myself bound as above; and in token of the truth, with my own hand have subscribed the present schedule of my abjuration, and have recited it word by word. In Rome, at the Convent della Minerva, this twenty-second day of June 1633.

I, GALILEO GALILEI,
have abjured as above, with my own hand.

Galileo Galilei

The Church's victory was complete, at least for a while. Galileo had long realized that his book would be banned, but he had hoped to escape imprisonment. That he did not was a final humiliation and betrayal. He completed his confession a broken man.

Above *A tapestry depicting Pope Urban VIII receiving a tribute from all nations.*

There exists a story, now well established in folklore, that Galileo, on rising from his knees after his abjuration, muttered under his breath *Eppure si muove!*, meaning "And yet it does move!" Perhaps this may be one of those sentences put into the mouths of great men after an historical event, but it cannot be true. Had Galileo uttered such words in plain hearing, or shown an obvious sign of defiance, he would have been consigned to the dungeons and would probably not have survived very long.

Galileo's humiliation and sentence did not end there. Immediately after the ceremony, copies of the sentence and the abjuration were sent to all the apostolic nuncios. The Inquisitor General in Florence was ordered to read both documents publicly in the hall of the Inquisition, and to serve notices to attend on all Galileo's disciples and adherents, as well as all public professors. Everyone, supporters and opponents alike, were made to participate in his disgrace. However, none of the decrees, orders, or official documents relating to the trial of Galileo were ever officially ratified by the Pope, and his signature is nowhere to be found. Theologically, without the pontiff's signature the judgment was not officially infallible, but merely represented the fallible judgment of an assembly of cardinals. Ultimately, it seemed that Galileo had not been persecuted by Pope Urban VIII, the infallible Vicar of Christ, but by the real persona beneath the vestige of pontiff: his old friend Maffeo Barberini, a pompous, mean, irascible, and vain man with a wounded pride.

It was decreed that all Galileo's works, even those he may publish in the future, were to be banned and placed on the Vatican's index of heretical works. In this way, Galileo was silenced. Some believed that the Jesuits had been the real culprits who had plotted for years against Galileo and eventually succeeded in having him tried for heresy. Later it was said that had Galileo not fallen out with them, but had he flattered the Jesuits and retained their favors, he could have been able to promote his science, written anything he liked, and stood in renown before the world.

Given the magnitude of his victory, Pope Urban wished to mitigate the severity of the sentence. Perpetual imprisonment was immediately commuted to house arrest in the Villa Medici, in the pleasant gardens of Trinita' del Monte, where Galileo, years

Left *A portrait of Grand Duke Ferdinand de' Medici, who had been Galileo's pupil as a child and who became his protector through the last stages of his life and during the trial.*

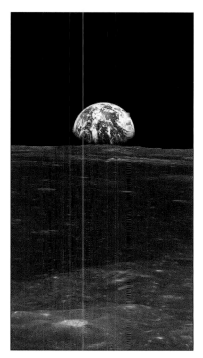

Above *Earth-rise seen from the moon: science would prove Galileo correct.*

before, had shown a group of cardinals the satellites of Jupiter. "I conducted him there," Ambassador Niccolini wrote, "and there he remains awaiting the clemency of his Holiness. He and Cardinal Barberini do not think it fit to grant a free pardon, but they will at all events allow him to go to the Archbishop's at Siena, or else to some convent in that city. I hear that Galileo has been much cast down at the punishment, of which he has just been informed. As to the book, he did not care for its being prohibited, and indeed had foreseen that it would be so." Shortly after, Niccolini saw the pope again, and pleaded that, as soon as the plague abated, Galileo might be allowed to live a prisoner in his own house at Arcetri. His Holiness replied that it was too soon to discuss further commutation as yet, and that allowing Galileo to be a prisoner in the Archbishop's palace instead of in a convent was proof of great leniency.

Suor Maria Celeste wrote: "I feared something must have happened, and importuned Signor Geri to tell me; but what I hear from him of the resolution they have taken concerning you and your book gives me extremest pain, not having expected such a result. Dearest lord and father, now is the time for the exercise of that wisdom with which God has endowed you. Thus you will bear these blows with that fortitude of soul which religion, your age, and your profession alike demand." Later she wrote, "I wish, that I could describe the rejoicing of all the mothers and sisters on hearing of your happy arrival at Siena. It was indeed most extraordinary! On hearing the news, Mother Abbess and many of the nuns ran to me, embracing me and weeping for joy and tenderness." Despite his pleasant surroundings Galileo was despondent, "My name is erased from the book of the living," he wrote. "Nay," came Suor Maria Celeste's reply, "say not that your name is struck out for it is not so; neither in the greater part of letter, the world nor in your own country. Indeed it seems to me that if for a brief moment your name and fame were clouded, they are now restored to greater brightness."

Galileo asked Niccolini if there was any way he could use his influence with the Grand Duke to talk to the pope to allow him to return home to Florence. In his depressed state, he missed his daughter and Maria Celeste's letters came almost daily. She told

Above *Villa Medici at Santa Trinita'
dei Monte in Rome, the residence of
Ambassador Niccolini, the diplomatic envoy
of the Grand Duke of Tuscany. It is here that
Galileo first returned to after his trial.*

Above *Cloistered life was miserable and hard. Galileo's beloved daughter, Suor Maria Celeste, died prematurely, depriving Galileo of a vital support.*

him of the fruit and the wine which were sold by the convent, and of the vines injured by hail, thieves plundering the garden, and of a terrible storm destroying one end of the roof. She explained that there had been few plums, and with the money from the lemons, two lire, she had had three masses said for her father. "There are beans in the garden waiting for you to gather them. Your tower is lamenting your long absence," she added.

Suor Maria Celeste, afflicted by continual ill health, by working nightly in the infirmary, and arduous daily chores, seemed to have a premonition of her death. When news reached her that her father's prison had been changed to Arcetri, and that he would shortly set out on his return, she had not enough life in her to be glad, "I do not think," she wrote in December 1633, "that I shall live to see that hour. Yet may God grant it, if it be for the best."

Her last prayer was granted, and she survived three more months. Galileo described the end, "I stayed five months at Siena in the house of the Archbishop after which my prison was changed to confinement to my own house, that little villa a mile from Florence, with strict injunctions not to entertain friends, nor to allow the assembling of many at a time. Here I lived on very quietly, frequently paying visits to the neighboring convent, where I had two daughters who were nuns, and whom I loved dearly; but the eldest in particular, who was a woman of exquisite mind, singular goodness, and most tenderly attached to me. She had suffered much from ill-health during my absence, but had not paid much attention to herself. At length dysentery came on, and she died after six days' illness, leaving me in deep affliction." Suor Maria Celeste died on the April 1, 1634. She was only thirty-three-years-old.

Below *View of the Santa Croce Church and Monastery where Suor Maria Celeste was buried.*

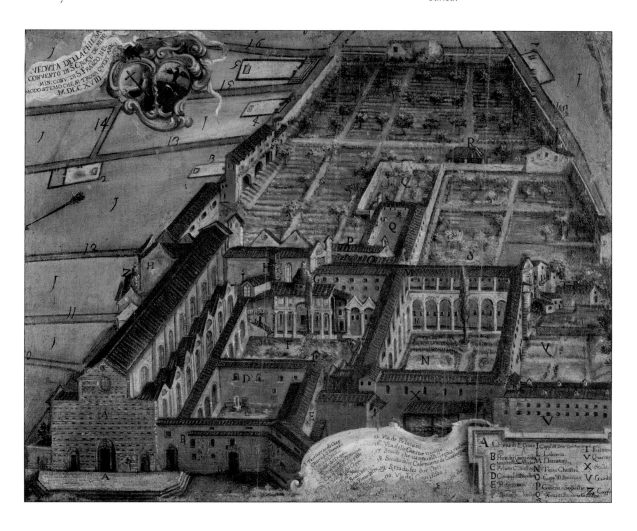

Above *The astronomical observatory at Arcetri on the outskirts of Florence that was built in Galileo's honor.*

"From this answer, I may conclude that my present prison will only be changed for that last narrow dwelling common to us all."
Galileo

It must have seemed that all Galileo cared for had gone or was now closed to him. The conditions placed on his house arrest were strict, and he lived in almost perpetual solitude. He could not complain or ask for more leniency, because he had been told that the result of any further application would be his instant consignment to the dungeons of the Inquisition in Rome (*below left*). Galileo's health declined so rapidly after the death of his daughter Maria Celeste, that it seemed to him, and all those who saw him, that he was destined to follow her quite quickly, and he believed himself to be dying. In time, however, his interest in work returned and his mind moved to the problems of motion. Work became for Galileo more than just mere consolation, almost a necessity , although he was aware that his mental powers were failing, "My restless brain goes grinding on," he wrote in a letter, "in a way that causes great waste of time; because the thought which comes last into my head in respect of some novelty, drives out all that had been there before."

He was working on his *Dialogues on Motion*, wishing, as he told friends, that the world should see the last of his labors before his time of final departure. As he worked and wrote, thoughts and scientific ideas came thick and fast upon him, so that his work increased, while each day shortened his span of life. *The Dialogues on Motion* were completed in the summer of 1636, and, because his books were banned in Italy and all catholic nations, he managed to have the manuscript taken to Holland for publication. The Inquisition-condemned *Dialogue* had already been translated into English much to the gratification of its author, and to the rage and mortification of the Jesuits. Galileo dedicated his last work to his old pupil François de Noailles (1584–1645), who while French ambassador in Rome, had used his own influence and that of another of Galileo's students, the Duke of Peiresc (dates uncertain), to mitigate Galileo's continued punishment. No sooner were the *Dialogues on Motion* written, that his busy brain began to form new projects. "If I live," he wrote, "I intend to put in order a series of natural and mathematical problems, which I trust will be as curious as they are novel."

In reality, there was nothing Galileo could do to rise above the bitterness he felt about the Church. Two examples of the letters he wrote during this dark period show how much he hated the Jesuits for what they had done to him. The first is a letter to the Duke of Peiresc, dated February 21, 1636 (*right*):

I have said, my lord, that I hope for no alleviation; and this is because I have committed no crime. If I had erred, I might hope to obtain grace and pardon; for the transgressions of the subject are the means by which the prince finds occasion for the exercise of mercy and indulgence. Wherefore, when a man is wrongly condemned to punishment, it becomes necessary for his judges to use the greater severity, in order to cover their own misapplication of the law. This afflicts me less than people may think possible; for I possess two sources of perpetual comfort: first, that in my writings cannot be found the faintest shadow of irreverence toward Holy Church, and secondly, the testimony of my own conscience, which I myself alone know thoroughly, besides God in heaven. And He knoweth that in this cause for which I suffer, though many men might have spoken more learnedly, none, not even the ancient fathers, have spoken with more piety, nor with greater zeal for Holy Church, than I. Could all the frauds, the calumnies, the stratagems, the deceits, which were made use of at Rome eighteen years ago for the purpose of imposing upon the supreme authority, could all these, I say, be brought to light, their only effect would be to enhance the purity and uprightness of my intentions. But you, having read my works, will have seen how they justify my assertions of sincerity, and you will have understood the true cause for which, under the mask of religion, I have been persecuted, and which now continually assails me and crosses my path; so that no help can come to me from without, nor can I myself undertake my own defense. For all the Inquisitors have received express orders to allow neither the reprinting of such works of mine as were published many years ago, nor to grant a license to any fresh work which I may desire to publish. Thus I am not only forced to keep silence towards those who strive to distort my doctrine and make my ignorance manifest, but I must also bear the insults and the contempt and the bitter taunts of men more ignorant than myself, without proffering a word in my own defense. My heart thanks you better than my words can express, for the most pious and humane office which you have undertaken on my behalf. May the Lord reward you for your kindness.

Galileo Galilei

He wrote a letter to Ladislaus, King of Poland, in 1637 (*left*).

Tragedy seemed to be stalking him at every turn. His sister-in-law, Clara Galilei, with her three daughters and one son, came to live with him; but they all perished in the plague shortly after their arrival. His eyesight was also failing.

Galileo made a last celestial discovery: the nodding and swaying of the moon's disc, a movement know as libration. "I do go on speculating," he wrote to a fellow scientist, "but to the great prejudice of my health; for thinking, joined to various other molestations, destroys my sleep, and increases the melancholy of my nights; while the pleasure which I have taken hitherto in making observations on new phenomena is almost entirely gone. I have observed a most marvelous appearance on the surface of the moon. Though she has been looked at such millions of times by such millions of men, I do not find that any have observed the slightest alteration in her surface, but that exactly the same side has always been supposed to be represented to our eyes. Now I find that such is not the case, but on the contrary that she changes her aspect, as one who, having his full face turned towards us, should move it sideways, first to the right and then to the left, or should raise and then lower it, and lastly incline it first to the right, then to the left shoulder. All these changes I see in the moon; and the large, anciently known spots which are seen in her face, may help to make evident the truth of what I say. Add to this a second marvel, which is that these three mutations have their three several periods; the first daily, the second monthly, the third yearly. Now what connection does your Reverence think these three lunar

I send your Majesty three lenses, according to the command which I received in your most gracious letter. I have endeavored to the utmost of my ability to serve your Majesty well in this matter; but I am in prison here, and have been for the last three years, by order of the Holy Office, for having printed the Dialogue on the Ptolemaic and Copernican systems; though I had the license of the Holy Office, that is, of the Master of the Sacred Palace of Rome. I know that some copies of the said books have penetrated your Majesty's dominions; and your Majesty and such of your subjects as call themselves scientific men may have judged whether or not it be true that my book contains doctrine more scandalous, more detestable, and more pernicious than is to be found in the writings of Luther, Calvin, and all other heresiarchs put together! Nevertheless, the Pope's mind has been so strongly imbued with this idea, that the book has been prohibited, and I put to utter shame, and condemned to imprisonment during his Holiness's pleasure, which will be perpetual. But where doth passion transport me? Let me go back to the lenses.

Galileo Galilei

periods may have with the daily, monthly, and annual movement of the sea? which is ruled over by the moon, by the consent of all."

This was the last of his long list of discoveries. His sight decayed rapidly.

"I have been in my bed for five weeks," he wrote to his faithful friend Elia Diodati (1576–1661), while there still remained a slight hope that the blindness might not be permanent, "oppressed with weakness and other infirmities from which my age, seventy four years, permits me not to hope

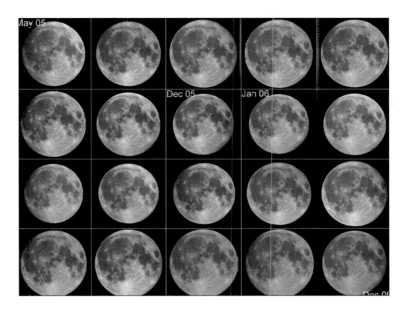

release. Added to this the sight of my right eye that eye whose labors (I dare say it) have had such glorious results is forever lost. That of the left, which was and is imperfect, is rendered null by a continual weeping." "Alas!" he wrote again to the same friend a few months later, "Your dear friend and servant Galileo has been for the last month hopelessly blind; so that this heaven, this earth, this universe, which I by my marvelous discoveries and clear demonstrations had enlarged a hundred thousand times beyond the belief of the wise men of bygone ages, henceforward for me is shrunk into such a small space as is filled by my own bodily sensations." Next to his bed, Galileo kept his father's lute.

Above *Galileo's lost discovery was the nodding and swaying of the moon's disk, a movement known as libration.*

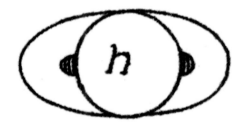

CHAPTER EIGHT

A SHRUNKEN UNIVERSE

Under house arrest and almost completely blind, Galileo's last
few years were lonely. He was not permitted to either visit or
receive friends by decree of the Holy Inquisition. After 1638,
it appears he never again left the confines of his home. The
universe he had opened to others through his observations now
closed in on him, and he died in January 1642. His legacy lived
on, and even the Church, finally, admitted error,
centuries after he had been condemned.

The cataracts that increasingly clouded Galileo's eyes were left untreated, and after six months he became incurably blind. Only on a few occasions did Church orders allow him to leave his villa at Arcetri. Once, in the spring of 1638, he was permitted to visit a physician, but only on condition of not attempting to see or receive any of his friends during his outing. On another occasion, he could go to pay his respects to his friend the Count de Noailles in the town of Poggibonsi. Once or twice, he visited the Grand Duke at his Villa La Petraia in great secrecy, going from Arcetri in a curtained carriage in the early morning, and returning late at night. However, after 1638 it does not appear that he ever attempted to leave the confines of his home. If nothing else, increased feebleness kept him a prisoner.

Galileo felt his isolation bitterly. He frequently ended his letters with "from my prison in Arcetri," although occasionally he did receive visitors. The English philosopher Thomas Hobbes came, as did a young John Milton who was appalled to see what the Church had done to the great man.

The last scientific work of Galileo's life was a short treatise on the secondary light of the moon, a subject he had corresponded on with a professor in Padua, who said the moon was phosphorescent. Galileo disagreed. His blindness was a common topic in most of the letters he wrote during this period. Writing to Prince Leopold dé Medici on March 13, 1640, he said, "I am obliged to have recourse to other hands and other pens than mine since

Below *A drawing of the Copernican model of the universe.*

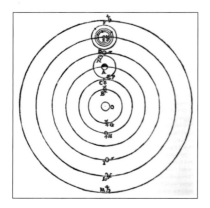

Above Villa La Petraia, the country
residence of the Medici Grand Dukes.

my sad loss of sight. This, of course, occasions great loss of time, particularly now that my memory is impaired by advanced age; so that in placing my thoughts on paper, many and many a time I am forced to have the foregoing sentences read to me before I can tell what ought to follow; else I should repeat the same thing over and over. Your Highness may take my word for it: that between using one's own eyes and hands and those of others there is as great a difference as in playing chess with one's eyes open or blindfolded."

In 1639, the Vatican allowed a seventeen-year-old youth called Vincenzo Viviani (1622–1703) to live with him. Later, one of his students, Evangelista Torricelli (1608–1647), came to live in the house also. Together they dealt with his correspondence and took down his dictations. Galileo had the idea of adding two more chapters, or conversations, to his *Dialogues on Motion*, with the last of these devoted to a discussion on the force of percussion. More often than not Galileo was too ill to work, and his assistants angled his bed so that he could feel the cool breeze and hear birdsong, and sometimes the taunting church bells of nearby Florence. Even his father's lute went untouched as he remembered the sunny days and madrigals of youth. With much work left undone and new avenues of science opening before him, he continued to suffer and feel his universe close in on him. As he lay dying, Galileo must have known he would be proven right, but if this provided him solace we do not know. After two months of struggling with ill health, he died on the evening of January 8, 1642, with Viviani and Torricelli by his side.

The news spread across Italy and the world: "Today news has come of the loss of Signor Galilei, which touches not just Florence, but the whole world and our whole century that from this divine man has received more splendor than from almost all the other ordinary philosophers. Now, envy ceasing, the sublimity of that intellect will begin to be known which will serve all posterity as a guide in the search for truth." For many, the word "truth" would be his epitaph. His son, Vincenzo, once said that his father knew the beauty of truth.

In his will, made in 1638, Galileo asked that his body be buried in the family vault at Santa Croce Church in Florence. His son Vincenzo was the sole heir; to his daughter, Sister Arcangela, he

gave an annuity of twenty-five crowns; to his nephews Alberto and Cosimo, then living in Munich, he bequeathed a thousand crowns. This bequest, however, was revoked in a codicil added the same year. He willed that any of his descendants who entered a religious order would receive nothing from his estate. After his death, not only was Galileo's power in making a will disputed because of his poor physical condition, but the propriety of laying his body in consecrated ground was also questioned. Eventually lawyers upheld the validity of his will and argued successfully that his body could be buried in a church.

His body was brought from Arcetri to the church of Santa Croce in Florence, and preparations were made for a funeral that reflected the sense of loss felt by the Medici Court and shared by the whole city. The sum of three thousand crowns was quickly

Right *The Chapel del Noviziato in Santa Croce where Galileo was first buried.*

Below *Marble bust of Galileo with his telescope in hand.*

awarded to cover the expense of a marble mausoleum. But this, as other details, were instantly reported to the Holy Office in Rome, and Ambassador Niccolini received orders from the pope to tell the Grand Duke that if the funeral of Galileo were carried out as intended by the Medici court, it would prove most pernicious, and that the Grand Duke must remember that Galileo had caused scandal to all Christendom by his false and damnable doctrine when alive. Niccolini advised that the plan for a public funeral and oration, and the building of a mausoleum, should be curtailed, at least while the pope remained theologically in absolute charge of all consecrated ground. However, the people of Florence did not agree.

So powerful was the population's wish to honor Galileo that the Grand Duke felt obliged to bend a little toward his people, even if it implied going against the wishes of the pope. Although the plans for a public funeral and monument were set aside, Galileo's friends placed his remains in the Chapel Del Noviziato, at the end of the corridor leading from the south transept to the great sacristy of the Church of Santa Croce. Only a few attended his funeral. Out of respect to the pope, the Grand Duke remained in his villa in Pisa. In Santa Croce, in an obscure corner on the gospel side of the altar, Gelileo's body rested for almost a century along side that of his daughter Suor Maria Celeste. He was later reburied in the main body of the church in 1737, where a monument to him was finally erected. Pope Urban VIII, who had so tormented his former friend, lived only two years longer than Galileo, and to a degree the bitterness he felt about Galileo abated on his death. The pontiff's death was supposedly hastened by chagrin at the result of the First War of Castro—a struggle between the papacy and the dukes of Castro in central Italy. The

Right *The Hubble space telescope. This is the twenty-first-century version of Galileo's telescope and continues the nature of observation that exemplifies the theoretical models created in his time and thereafter.*

outcome was perceived as disgraceful to the papacy, as it had been unable to enforce its will by military dominance. The costs incurred by the city of Rome to finance the war had been very high, and as a consequence Urban VIII became immensely unpopular. He was known as the "taxing pope." On his death, the bust of Urban that lay beside the Conservator's Palace on the Capitoline Hill was rapidly destroyed by an enraged crowd. The pope died at 10.45 in the morning, and it was said that by mid-day the bust had been shattered, and only a quick-thinking priest was able to save the sculpture of Urban that belonged to the Jesuits from a similar fate.

Throughout the seventeenth century, improvements continued to be made in telescopes and more and more was revealed of the night sky. Soon there was no doubt left that what Galileo had observed was indeed real. Eventually it was realized that strange Saturn—which Galileo thought could possibly be a triple planet—was in fact a ringed world. The Inquisition lifted its ban on reprinting Galileo's works in 1718, and by this time it was obvious to all that Copernicus had been correct, when permission was given to produce an edition of his works—albeit excluding the condemned *Dialogue*—in Florence. In 1741, Pope Benedict XIV authorized the publication of an edition of Galileo's complete scientific works that included a slightly censored version of the *Dialogue*. In 1758, the Church was no longer able to hold back the tide of scientific progress, and the general prohibition against works advocating Copernican theories was removed from the Index of Prohibited Books, although the specific ban on uncensored versions of the *Dialogue* and of Copernicus's *De Revolutionibus* remained. The last traces of official opposition by the Church disappeared in 1835, when these works were finally eliminated from its Index. By then the scientific age that Galileo had started was underway, and even the Vatican had built an observatory and employed astronomers.

But what of Galileo? It was as recently as 1939 that Pope Pius XII, in his first speech to the Pontifical Academy of Sciences made within a few months of his election to the papacy, described Galileo as being among the "most audacious heroes of research ...not afraid of the stumbling blocks and the risks on the way, nor fearful of the funereal monuments." A papal advisor of the time also said, "Pius

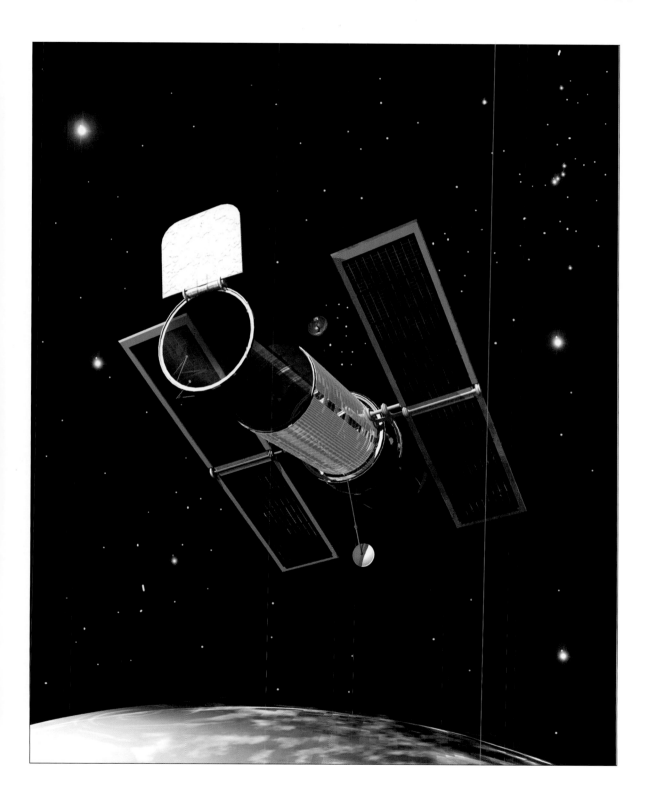

Above *The WEBB telescope that will supercede the Hubble telescope in 2011. Galileo's entire work is justified in the work now carried out by these powerful telescopes.*

XII was very careful not to close any doors (to science) prematurely. He was energetic on this point and regretted that in the case of Galileo."

The Galileo affair is the story of many tragedies. That of an old, blind prisoner, of a nation's scientific decline, and of the Church's estrangement from science. In February 1990, in a speech delivered at the University La Sapienza in Rome, Cardinal Ratzinger—later to become Pope Benedict XVI—summarized the Galileo affair as "a symptomatic case that permits us to see how deep the self-doubt of the modern age, of science and technology goes today."

Later, in October 1992, Pope John Paul II, speaking in the hall of the Apostolic Palace in the Vatican, expressed regret for how the affair was handled, and, three-hundred and fifty-nine years after the judgment against Galileo, officially conceded that the Earth was not stationary, as the result of a recent study conducted by the Pontifical Council for Culture confirmed. This was not an apology, merely an "admission of error."

In March 2008, a statue of Galileo was erected inside the Vatican's walls and later the same year during events to mark the four-hundredth anniversary of Galileo's earliest telescopic observations, Pope Benedict XVI praised his contributions to astronomy. A Vatican scholar said Galileo "lovingly cultivated his faith and his profound religious conviction and was a man of faith who saw nature as a book authored by God." Indeed, the Galileo anniversary appears to be giving the Vatican new impetus to finally put the matter to rest. In so doing, Vatican officials emphasized Galileo's strong faith as well as his science, to show that the two are not mutually exclusive. However, this may be too late. Another tragedy of the Galileo affair is that while science and technology have changed the world, the Catholic Church has followed many paces behind in the advance of civilization. Perhaps, on his anniversary, things will begin to change.

Pope Benedict XVI paid Galileo tribute saying he and other scientists had helped the faithful better understand and "contemplate with gratitude the Lord's works." It has even been suggested that Galileo should be named the patron of the dialogue between faith and reason. What Galileo himself would have made of this suggestion we can only guess.

When Galileo's body was exhumed and reburied in 1737, a vertebra, three fingers, and a tooth were removed. The finger that pointed the way to the stars is on display and can be seen in the Institute and Museum of the History of Science of Florence.

Postscript

SCIENCE RISING

A biography of Galileo Galilei takes the reader on a journey through space and time. The astronomical revolution heralded by Copernicus that was confirmed by Galileo's observations of the heavens, led to discoveries by Johannes Kepler, René Descartes, and Isaac Newton. Albert Einstein regarded the scientific reasoning by Galileo as one of the most important achievements in the history of human thought. Galileo's work marks the real beginning of physics and his legacy is still with us today.

G alileo believed that his most important scientific contribution was not the astronomical discoveries he made, nor his defense of Copernicus, but rather his application of mathematics to the study of motion, begun with his observation of the sway of a church lamp and the dropping of balls of different weights from the leaning tower of Pisa. Earlier philosophers, commencing with Aristotle, concerned themselves with the causes of motion, but Galileo, typically, looked at the problem from another angle. He concentrated on measurement, and departed from vague unscientific notions of "essences" and "natural places." What mattered to him were speed, time, distance, and acceleration, and the laws that bound them. Fundamental to him was science—its power and all its possibilities. His studies of projectiles and free-falling bodies brought him very close to the complete formulation of the laws of inertia and acceleration that were later deduced by Isaac Newton. Newton said that he was able to see far because he had "stood on the shoulders of giants." None were more gigantic than Galileo.

It was no surprise that the next major advances, by René Descartes and Isaac Newton, were made in the Protestant north, for what had happened to Galileo had frightened Italian scientists and the nation itself. Both scientifically and in terms of political influence, Italy began a long decline that lasted centuries. Perhaps it is too simplistic to blame this on the Galileo affair, but certainly Maffeo Barberini—the vengeful Pope Urban VIII—and the revenge of the Jesuits were to cast negative influence in the course of history.

Below *The Galileo spacecraft reached Jupiter's moon Io on December 7, 1995. Galileo found the volcanic moon to have a huge iron core that takes up half its diameter.*

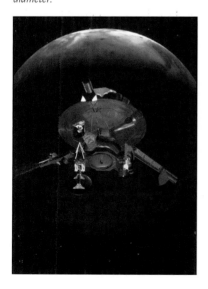

Above *The Galileo spacecraft orbiting Jupiter. The planet's moon, Io, is shown as the crescent to the left of Galileo. The distant sun is visible between Io and the spacecraft's long magnetometer. Jupiter is to the right.*

Above *This artist's rendering shows a view of our own Milky Way Galaxy and its central bar as it might appear if viewed from above. Astronomers have concluded that our galaxy harbors a stellar bar.*

Galileo believed that experiments and observations were the cornerstones of truth. He separated science from philosophy. "Philosophy," he wrote, "is written in this grand book the universe. But the book cannot be understood unless one first learns to comprehend the language and to read the alphabet in which it is composed. It is written in the language of mathematics, and its characters are triangles, circles, and other geometric figures, without which it is humanly impossible to understand a single word of it." Every experiment that a scientist performs today is

Galileo's legacy. With complete justification, Galileo has been called the father of modern physics and the father of modern science. Albert Einstein said, "The discovery and use of scientific reasoning by Galileo was one of the most important achievements in the history of human thought and marks the real beginning of physics." Galileo knew that science and its discoveries would provoke a profound shift in consciousness and in the way we understand and relate to our place in the universe. He understood that such shifts would cause a crisis, and when we face such moments again in the future we should remember Galileo.

Above *This spectacular image taken by the Spitzer wide-area Infrared Extragalactic Legacy Project, has captured the bright blue sources of light that are hot stars in our own Milky Way, ranging anywhere between three to sixty times the mass of our sun. The fainter green spots are cooler stars and galaxies beyond the Milky Way whose light is dominated by older stellar populations.*

Galileo was not the first to turn a telescope toward heaven, but his skill made him the most important scientist to do so, and his ability in making telescopes enabled him to see further into space than others, and beyond the limited boundaries of the Catholic Church. When the Apollo astronauts sped across the lunar landscape looking for a place to land, they were following Galileo's path. When we count sunspots to understand the rhythmic behavior of our local star, we are emulating Galileo. When a space probe called Galileo penetrates Jupiter's system and sweeps past the tiny satellites he saw first, it pays him homage. What is Galileo's legacy? The universe around you.

GALILEO GALILEI, 1564-1642 CHRONOLOGY

Right *Galileo Galilei and Vincenzo Viviani discussing astronomy.*

1564

Galileo Galilei is born in Pisa on February 15, the son of Vincenzo Galilei (1520–1591) and Giulia Ammannati (1538–1620). Galileo's father, Vincenzo, was a composer, a music theorist, and a seminal figure in the musical life of the late Renaissance. In September 1573, Galileo begins to attend the classes given by a local teacher, Muzio Tedaldi. The following year, his father interns him in the Vallombrosa Monastery on the outskirts of Florence. He studies at Vallombrosa until 1578, when Vincenzo Galilei is employed by the Medici Grand Duke and the family moves to Florence. Vincenzo's discoveries in acoustics and in the physics of vibrating strings influence deeply Galileo. Vincenzo directs his son away from the pure mathematical formulas followed at the time, toward experimentation and observation, with results explained by empiricist mathematics.

1583–1587

Galileo begins to study mathematics at the University of Pisa under the tutor Ostilio Ricci. In 1585, he writes some theorems on the centre of gravity of solids. In that same year, he leaves the university and returns to Florence where he works as a private tutor of mathematics. In 1586, Galileo writes a treatise titled *La Bilancetta* (The Little Balance), that outlines the procedures for the construction of a balance. In 1587, he seeks the Chair of Mathematics of Bologna University, one of the most prestigious academic posts in Europe, but loses to a rival. Later that year, he gives a lecture in the Accademia in Florence on the size and location of Dante's Inferno.

1589–1594

In 1589, Galileo writes the treatise *De Motu* (On Motion), and bids successfully for the Chair of Mathematics at the University of Pisa. In 1591, Galileo's father, Vincenzo Galilei, dies and is buried at the Santa Croce Church in Florence. In 1592, he bids for the Chair of Mathematics at the University of Padua, a more prestigious university than Pisa, and is successful. He will teach at Padua until 1610. In the Padua years, he conducts studies and experiments in mechanics, builds the thermoscope, and invents and builds the geometric and military compass. In 1594, he patents a water-lifting machine.

1595–1616

In 1609, while still teaching mathematics at Padua, Galileo develops the telescope, with which he performs the observations that lead him to the discovery of the phases of the moon and also to the discovery of Jupiter's moons, still named collectively in his name. In 1610, he is appointed mathematician and philosopher to the Grand Duke of Tuscany. He studies the peculiar

Above *The leaning tower of Pisa from which Galileo supposedly dropped balls of different weights and sizes in order to find out if they all fell at the same speed.*

Below *An engraving showing Galileo's pendulum clock.*

Above *Galileo shows his telescope to the Doge of Venice.*

appearances of Saturn and observed the phases of Venus. In 1611, he goes to Rome, where he is invited to join the prestigious Accademia dei Lincei and observes sunspot activity more accurately. In 1612, Church opposition arises to Copernican theories, which Galileo supports. In 1614, from the pulpit of Santa Maria Novella, Father Tommaso Caccini (1574-1648) denounces Galileo's opinions on the motion of the Earth, judging them dangerous and close to heresy. Galileo goes to Rome to defend himself against these accusations. However, in 1616, Cardinal Roberto Bellarmine (1542-1621) personally hands Galileo an admonition enjoining him to neither advocate nor teach Copernican astronomy, because it is contrary to the accepted understanding of the Holy Scriptures.

1617-1632

In 1622, Galileo writes the *Saggiatore* (The Assayer), which is approved by the Vatican. He also develops the microscope which he shows in both Florence and Rome. In 1630, he returns to Rome to apply for a license to print the *Dialogo dei Massimi Sistemi* (Dialogue on the Great World Systems). The Vatican approves the text of the book and gives him the license. Thus, the treatise is published in Florence in 1632, amid an outbreak of the bubonic plague. But in October of that year, Galileo is ordered to appear before the Holy Office in Rome. His old friend and supporter, Maffeo Barberini, now Pope Urban VIII, takes issue with the views of the Church as represented in the *Dialogue*, and is determined to destroy the scientist. The Vatican court issues a sentence of condemnation

Above *Our solar system.*

and forces Galileo to abjure publicly. He is confined in Siena and eventually, in December 1633, he is allowed to retire to his villa in Arcetri under house arrest, despite the protection of the Medici Grand Duke.

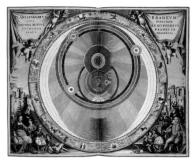

1633-1642

In 1634, Galileo is deprived of the support of his beloved daughter, Suor Maria Celeste (1600-1634), who dies prematurely. The Church continues to forbid him to go anywhere or for anyone to visit him. In 1638, when he is almost totally blind, the *Discorsi e Dimostrazioni Intorno a Due Nuove Scienze* (Discourses and Demonstrations on Two New Sciences) is published in Leiden. Galileo died in Arcetri on January 8, 1642. He is buried close to his father, daughters Arcangela and Maria Celeste, at Santa Croce Church in Florence.

Above *Tycho Brahe's planetary system, in a 1660 version that incorporates Galileo's discovery of four moons shown to its right. In the center is the Earth, encircled by the orbits of Mercury and Venus, and further out, those of Mars, Jupiter, and Saturn. An alternative position for the sun and its attendant satellites is shown in the lower half of the figure. Seated at the bottom right is Tycho Brahe, measuring a globe.*

1642-Present

The Galileo legacy lives on. The astronomical revolution heralded by Copernicus that was confirmed by Galileo's observations of the heavens, led to discoveries by Johannes Kepler, René Descartes, and Isaac Newton. Albert Einstein regarded the scientific reasoning by Galileo as one of the most important achievements in the history of human thought. Galileo's work marks the real beginning of physics and his legacy is still with us today.

Right *The Galileo spacecraft orbiting Jupiter. The planet's moon, Io, is shown as the crescent to the left of Galileo. The distant sun is visible between Io and the spacecraft's long magnetometer. Jupiter is to the right.*

BIBLIOGRAPHICAL REFERENCE LIST

All quoted material from Suor Maria Celeste's letters to her father, from Ambassador Niccolini to the Grand Duke of Tuscany, the quoted passages from Galileo Galilei's writings and books, and the quoted passages from the Inquisition's trial have been paraphrased from Mary Allan-Olney's *The Private Life of Galileo*, published by Nichols and Noyes in 1870.

Other bibliographical reference material includes:

Drake, Stillman.
Galileo at Work.
New York: Dover Publications, 1995.

Drake, Stillman.
Galileo: A Very Short Introduction.
Oxford: Oxford University Press, 2001.

Galilei, Galileo.
Sidereus Nuncius.
Chicago: Chicago University Press, 1989.

Feldhay, Rivka.
Galileo and the Church.
Cambridge: Cambridge University Press, 1995.

Koestler, Arthur.
The Sleepwalkers.
New York: Viking Penguin, 1989.

Redondi, Pietro.
Galileo: Heretic.
Princeton: Princeton University Press, 1989.

Reeves, Eileen.
Galileo's Glassworks.
Harvard: Harvard University Press, 2008.

Reston, James, Jr.
Galileo: A Life.
Knoxville: Beard Books, 2000.

Sobel, Dava.
Galileo's Daughter: A Historical Memoir of Science, Faith, and Love.
New York: Penguin, 2000.

FURTHER READING LIST

Dra ke, Stillman.
Galileo at Work.
New York: Dover Publications,
1995.

Drake, Stillman.
Galileo: A Very Short Introduction.
Oxford: Oxford University
Press, 2001.

Drake, Stillman.
Discoveries and Opinions of Galileo.
New York: Anchor Books, 1957.

Feldhay, Rivka.
Galileo and the Church.
Cambridge: Cambridge
University Press, 1995.

Fermi, Laura and Gilberto
Bernardini.
*Galileo and the Scientific
Revolution.*
New York: Dover Publications,
2003.

Galilei, Galileo and Maurice A
Finocchiaro.
The Essential Galileo.
Indianapolis: Hackett
Publishing, 2008.

Galilei, Galileo.
Sidereus Nuncius.
Chicago: Chicago University
Press, 1989.

Koestler, Arthur.
The Sleepwalkers.
New York: Viking Penguin,
1989.

Ness, Atlae,
trans. by S. Anderson,
*Galileo Galilei, When the World
Stood Still.*
New York: Springer, 2005.

Redondi, Pietro.
Galileo: Heretic.
Princeton: Princeton University
Press, 1989.

Reeves, Eileen.
Galileo's Glassworks.
Harvard: Harvard University
Press, 2008.

Reston, James, Jr.
Galileo: A Life.
Knoxville: Beard Books, 2000.

Sobel, Dava.
*Galileo's Daughter: A Historical
Memoir of Science,
Faith, and Love.*
New York: Penguin, 2000.

OTHER BOOKS BY THE AUTHOR

The Moon: A Biography
London: Headline Book
Publishing, 2001

The Sun: A Biography
Hoboken: John Wiley
and Sons, 2005

AUTHOR'S ACKNOWLEDGMENTS

I must thank, Al Zuckerman, Meredith Hale, Bob Diforio, Manuela Roosevelt, and Stefan Morris. My grateful thanks are also due to my wife, Jill.

David Whitehouse,
Hampshire, England,
March 2009

SPRINGWOOD'S ACKNOWLEDGMENTS

Springwood would like to thank Jot Singh, Liisu Carlson, and Priya Hemenway for their wonderful editorial assistance; Carolyn Castor for a rushed but perfect index, and Katja Lehmann of Archivio Scala for working with us.

ART ACKNOWLEDGMENTS

JACKET COVER
Cover art: Portrait of Galileo Galilei by Ottavio Leoni (c. 1578-1642), Biblioteca

Marucelliana, Florence, ©
Photo Scala Florence

Florence, Santa Croce Museum, © 1990 Photo Scala, Florence/ Fondo Edifici di Culto, Ministero dell'Interno

Page 28: Flemish armillery sphere 1562, © Adler Planetarium, Chicago.

Engraving of Galileo's pendulum clock, Biblioteca Nazionale Centrale, Florence

Page 30: Dante's Inferno by Sandro Botticelli, © Galleria degli Uffizi, Florence

CHAPTER TWO

Page 32: Computer Generated Image of feather and lead ball, © Springwood and Moon*Runner* Design

Page 34: Astronomer with astrolabe, Edimedia, Paris

Page 35: Computer Generated Image of Commander David Scott repeating Galileo's experiment on the moon, © Springwood and Moon*Runner* Design

Page 37: Computer Generated Image of balls of different weights thrown from the Tower of Pisa, © Springwood and Moon*Runner* Design

Page 38: Computer Generated Image of rolling inclines, © Springwood and Moon*Runner* Design

Page 39: Portrait of Don Giovanni de' Medici. Cerreto

Guidi, Medici Villa, © 1990. Photo Scala, Florence, courtesy of the Ministero Beni e Att. Culturali

Page 40: Portrait of Tycho Brahe, Royal Library, Copenhagen

Page 41: Portrait of Nicolas Copernicus, Detlev van Ravenswaay, © Science Photo Library

Page 43: Etching by Jan Luyken, Moravska Gallery, Brno

Page 44: Cardinal Bellarmine by Andrea Pozzo (1642–1709). Rome, Church of Sant'Ignazio, © Photo Scala, Florence/ Fondo Edifici di Culto, Min. dell'Interno

Page 45: Painting of Doge's Palace and St. Mark's Square by Canaletto (1697–1768), Galleria degli Uffizi, Florence, © Photo Scala Florence

Page 46: Portrait of Ptolemy from a fifteenth-century manuscript, Biblioteca Marciana, Florence

Page 47: Woodcut depicting Massahalla, Public Domain

Page 48: Galileo Galilei's thermoscope, Florence, Museo della Scienza, © 1990, Photo Scala, Florence

Page 49: View of Padua University, Padua City Museum

Page 50: The world's systems represented as clocks, Johannes Zahn, *Specula Physico-Mathematico-Historica Notabilium ac Mirabilium Scientiarium*, Nürnberg, 1696

Page 51: Portrait of Johannes Kepler, Sternwarte Kremsmuster, Austria

Page 52: Tycho Brahe's Uraniborg Observatory, *Astronomias Instauratae Mechanica*, Wandesburg, 1598

Page 53: *Aritmetica* by Gregor Reisch (1467–1525), British Museum, London

Page 55: Chandra X-ray image of a supernova, © Nasa Images

Page 56: Computer Generated Image of the Ptolemaic system, © Springwood and Moon*Runner* Design

Page 57: Woodcut of Christian philosopher, Public Domain
Page 58: Bust of Pope Paul V by Giovanni Lorenzo Bernini (1598–1680), Roma. Galleria Borghese, © 2005 Photo Scala, Florence, courtesy of Ministero Beni e Att. Culturali

CHAPTER THREE

Page 62: Astrologer at work, Public Domain

Page 63: View of Florence, Springwood collection

Page 64: The optical principle of the telescope, Public Domain

Page 65: Portrait of Hans Lipperhey, Public Domain

Page 66: The world system determined from the geometry of solids, Johannes Kepler's *Hamonices Mundi Libri*, Linz, 1619

Page 67: Dudley Barnes

Page 151: Gaspero Martellini (1785-1850): An Experiment at the Accademia del Cimento, Florence, Tribuna di Galileo, © 1990 Photo Scala, Florence

Page 154: Title page of *Dialogo* by Galileo, 1632. Oxford, Science Archive, © 2005 Photo Scala, Florence/HIP

Page 158: Computer generated Image of Kepler's view of the universe, © Springwood and Moon*Runner* Design

Page 159: Luigi Sabatelli (1772-1850) The Old Age of Galileo, Florence, Tribuna di Galileo, © 1990 Photo Scala, Florence

Page 160: Benedetto Bonfigli (c. 1420-1496): Madonna della Misericordia, detail. Perugia, San Francesco (Oddi Chapel), © 1990 Photo Scala, Florence

Page 163: Computer Generated Image of five platonic solids, © Springwood and Moon*Runner* Design

Page 165: Joseph Nicholas Robert Fleury (1797-1890): Galileo Galilei devant le Saint Office au Vatican. 1847, Paris, Musée du Louvre, © 2008 White Images/Scala, Florence

Page 169: Joseph Clerian (1796-1842): Detail de Galilee devant le Tribunal de l'Inquisition, Aix-en-Provence, Musée Granet, © 2008 White Images/La Scala

Chapter Seven
Page 172: Moon, © Nasa Images

Page 174: Galileo during his trial, © Bridgeman Art Library, London

Page 175: Pierre Beauvallet (1801-1873), Représentation de Galileo Galilei dans la pièce "Galilee" de François Ponsard a la Comedie Française en 1867. © 2008 White Images/Scala, Florence

Page 178-179: Francisco de Goya (1746-1828), Scene of Inquisition, Madrid, Museo de la Real Academia de Bellas Artes de San Fernando, © 1995 Photo Scala, Florence

Page 181: Pedro Berruguete (1450-1504) Scenes from the life of Saint Dominic Guzman,

The Burning of the Books, Madrid, Museo del Prado, © 1990 Photo Scala, Florence

Page 182: Burning of heretics sentenced by the Inquisition, from History of the Inquisition, Cologne 1759, © 2005 Photo Ann Ronan/HIP/Scala, Florence

Page 184: Portrait of Galileo Galilei, British Library, London, © 2003 Photo Scala/HIP

Page 185: Barberini Factory (17th century): Series of tapestries on the life of Urban VIII, Vatican, Gallery of the Tapestries, © 1990 Photo Scala, Florence

Page 191: Barberini Factory (17th century): Series of tapestries on the life of Urban VIII, Vatican, Gallery of the Tapestries, © 1990 Photo Scala, Florence

Page 192: Justus Sustermans (1597–1681): Portrait of Ferdinand II de' Medici,

Florence, Galleria Palatina, © 1990 Photo Scala, Florence, courtesy of Ministero Beni e Att. Culturali

Page 193: Moonrise, © Nasa Images

Page 195: François Marius Granet (1775–1849): Trinita' dei Monti e Villa Medici, Paris, Musée du Louvre, © 1990 Photo Scala, Florence

Page 196: Cloister, Public Domain

Page 197: View of the Church and Monastery of Santa Croce, Florence, Santa Croce Museum, © 1990 Photo Scala, Florence/ Fondo Edifici di Culto, Ministero dell'Interno

Page 198: View of the astrophysical observatory at Arcetri, Florence, © 1990 Photo Scala, Florence

Page 201: Moons, © Nasa Images

CHAPTER EIGHT
Page 202: A drawing by Galileo, © Biblioteca Nazionale Centrale, Florence

Page 204: Copernican model of the universe, Public Domain

Page 205: Villa La Petraia, view of the garden toward

the villa, © 2005 Photo Scala, Florence, courtesy of the Ministero Beni e Att. Culturali

Page 207: Piazza Santa Croce, Florence, Museo di Forenze Com'Era, © 1990 Photo Scala, Florence

Page 208: Giovan Battista Foggini (1652–1725), Cenotaph of Galileo, Santa Croce, Florence, © 1990 Photo Scala,

Florence/ Fondo Edifici di Culto, Ministero dell'Interno

Page 209: View of the Novices Chapel, or Medici Chapel, Santa Croce, Florence © 1990 Photo Scala, Florence/ Fondo Edifici di Culto, Ministero dell'Interno

Page 211: Hubble space telescope, © Nasa Images

Page 212: Webb telescope, © Nasa Images

POSTSCRIPT

Page 214: Galileo spacecraft, © Nasa Images

Page 216: Galileo spacecraft, © Nasa Images

Page 217: Galileo spacecraft, © Nasa Images

Page 218: Milky Way, © Nasa Images

Page 219: Night sky, © Nasa Images

CHRONOLOGY

Page 220: Painting by Tito Lessi (1858–1917), © Florence, Museo della Scienza

Page 221: Computer Generated Image of balls of different weights thrown from the Tower of Pisa, © Springwood and Moon*Runner* Design

Page 221: Engraving of Galileo's pendulum clock, Biblioteca Nazionale Centrale, Florence

Page 222: Galileo shows his telescope to the Doge of Venice, Istituto e Museo della Storia della Scienza, Florence

Page 222: Computer Generated Image of our solar system, Springwood and Moon*Runner* Design

Page 223: Tycho Brahe's planetary system, Doppelmeier, 1742

Page 223: Galileo spacecraft, © Nasa Images

INDEX

Illustrations and documents will be designated by *fig* and *doc*, respectively, with their page numbers. The words *letter* and *quote* designate their use with page numbers.

A

B

C

G

J

M

Maculani, Vincenzo, 177, 180, 183

Magalotti, Fillipo, 166-67

Magini, Giovanni Antonio, 27, 40, 94

Maraffi, Luigi, 108, 111

Mare Imbrium (Sea of Rains), mountains east of, 86-87

Maria Celeste, Suor
on Archanela's illness, 151-52
birth of, 54
death of, 197, 198
enters convent, 8, 140, 142
on father being held in palace at Siena, 194
on father's before Inquisition, 180
and father's burial, 208
on father's ill health, 150, 152, 157, 160, 170
Galileo's close relationship with, 137, 148-49
on hardships at convent, 149-50, 152-53
the ill health of, 196-97
as not understanding father's situation, 177
on Vincenzo's behavior, 161

Marius, Simon, 69-70, 95, 97

Marzimedici, Alessandro, 121

Massimiliana, 150

Master of the Sacred Palace, 157

mathematics
Caccini calls for banishment of, 108

divine order in, 18*fig*, 26*fig*, 114*fig*, 132*fig*
hidden order in, 22, 25, 36
as philosophy to explain the universe, 145-46, 218-19
as replacement to Aristotelian rules, 19

Matti, Girolamo, 170

Maurice of Nassau, 65-67, 69

Maximilian I, 52

Mazzoleni, Marcantonio, 51

Mazzoni, Jacopo, 39

mechanical advancements, 50
military compass, 33, 51
ship design at Venice Arsenale, 44, 47
thermoscope, 11, 33, 47-48
water-lifting machine, 33, 48
see also physics; telescope

Mecury (planet), 106

Medicean stars/planets.
see Galilean moons

Medici, Cosimo de', I, 17

Medici, Cosimo de', II
appoints Galileo, 98
discusses planetary movements with Castelli, 122, 124-26
Galileo dedicates *Bodies of Water* to, 116
Galileo explains Venice gift to, 78-79, 81
and Galileo in Rome (1615-1616), 127, 129, 130, 134, 136-37
Galileo on his planetary observations to, 100, 103
Galileo seeks appointment to,

N

Neptune, 122

Netherlands, 65-69, 70, 75, 199

Newton, Isaac, 215, 216

Niccolini, Francesco, 162, 164, 166, 167
    advises Galileo to be compliant before Inquisition, 174, 177, 183
    audience with Urbane on Galileo, 164, 166, 171, 174, 176, 194
    defends Galileo in Rome, 162, 164, 166-68, 170-71, 174, 176

Galileo in home of, 170, 183, 195*fig*

and Galileo's funeral, 208

on Galileo's punishment by Inquisition, 185, 194

on Galileo's treatment by Inquisition, 180

on Galileo's view of God, 174

Nimrod, 31

Noailles, Francois de, 199, 204

Nori, Francesco, 114

O

"Old Discoverer," 84

On the new star in Ophiuchus's foot (Kepler), 54

On the revolutions of celestial spheres (Copernicus), 14, 42, 143, 210

On Motion (Galileo), 36

opera, 25

Ophiuchus constellation, 54, 57

Oreggi, Agostino, 135-36

Orion constellation, 87, 98*fig*

Orsini, Cardinal, 135

P

Padua, 77, 84, 97
    atmosphere of, 33, 40, 48, 50
    Galileo as outsider in, 39
    "stranger" with telescope visits, 75, 77

Paris, 73, 75

Paul III, 42

Paul V, 129
    appearance and character of, 103, 135
    Ballarmine as chief advisor to, 103
    death of, 145

T

U